# Motion Simulation and Mechanism Design Using SOLIDWORKS Motion 2017

**Kuang-Hua Chang, Ph.D.**

**School of Aerospace and Mechanical Engineering**
**The University of Oklahoma**
**Norman, OK**

**SDC**
**Publications**

**SDC Publications**
P.O. Box 1334
Mission, KS 66222
913-262-2664
www.SDCpublications.com
Publisher: Stephen Schroff

ISBN-13: 978-1-63057-082-8
ISBN-10: 1-63057-082-6


Printed and bound in the United States of America.

**Preface**

This book is written to help you become familiar with *SOLIDWORKS Motion* (also called *Motion* in this book), which is an add-in module of the *SOLIDWORKS* software. *Motion* supports users to conduct modeling and analysis (or simulation) of mechanisms in a virtual (computer) environment. Capabilities in *Motion* allow you to use solid models created in *SOLIDWORKS* to simulate and visualize mechanism motion and system performance. Using *Motion* early in the product development stage could prevent costly (and often painful) redesign due to design defects found in the physical testing phase. Therefore, in general, using *Motion* for support of design decision making contributes to a more cost effective, reliable, and efficient product design process.

This book covers basic concepts and frequently used commands required to advance readers from a novice to an intermediate level in using *SOLIDWORKS Motion*. Basic concepts discussed in this book include model generation, such as creating assembly mates for proper motion; carrying out simulation and animation; and visualizing simulation results, such as graphs and spreadsheet data. These concepts are introduced using simple yet realistic examples.

Verifying results obtained from the computer simulation is extremely important. One of the unique features of this book is the incorporation of theoretical discussions for kinematic and dynamic analyses in conjunction with the simulation results obtained using *Motion*. The purpose of the theoretical discussions lies in solely supporting the verification of simulation results, rather than providing an in-depth discussion on the subject of mechanism design. *SOLIDWORKS Motion* is not foolproof. It requires a certain level of experience and expertise to master the software. Before arriving at that level, it is critical for you to verify simulation results whenever possible. Verifying simulation results will increase your confidence in using the software and prevent you from being fooled (hopefully, only occasionally) by erroneous simulation results produced by the software.

Example files have been prepared for you to go through all lessons, including *SOLIDWORKS* parts and assemblies. Completed *Motion* simulation models are also provided for your references. You may want to start each lesson by reviewing the introduction and model sections and opening the assembly in *SOLIDWORKS Motion* to see the motion simulation, in hopes of gaining more understanding about the example problems. In addition, *Excel* spreadsheets that support the theoretical verifications of selected examples are also available. You may download all model files and *Excel* files from the web site of *SDC Publications* at:

http://www.sdcpublications.com/downloads/978-1-63057-082-8

This book is written following a project-based learning approach and is intentionally kept simple to help you learn *Motion*. Therefore, it is not intended in this book to include every single detail about *Motion*. For a complete reference of *Motion*, you may use on-line help in *SOLIDWORKS Motion*, or visit the web site of *SOLIDWORKS Corporation* at:

http://www.solidworks.com/

This book should serve self-learners well. If such describes you, you are expected to have basic Physics and Mathematics background, preferably a bachelor or associate degree in science or engineering. In addition, this book assumes that you are familiar with the basic concept and operation of *SOLIDWORKS* part and assembly modes. A self-learner should be able to complete all lessons in this book in about fifty hours. An investment of fifty hours should advance you from a novice to an intermediate level user. Not a bad investment at all!!

This book should also serve class instructors well. The book will most likely be used as a supplemental textbook for courses like *Mechanism Design*, *Rigid Body Dynamics*, *Computer-Aided Design*, *Computer-Aided Engineering*, *Design Principles*, or *Capstone Design*. This book should cover four to six weeks of class instruction, depending on how the courses are taught and the technical background of the students. Some of the exercise problems given at the end of the lessons may require significant effort for students to complete. The author strongly encourages instructors and/or teaching assistants to go through those exercises before assigning them to students.

Examples presented in this book have been verified using *SOLIDWORKS 2017 SP1.0*. If you are using a different *SOLIDWORKS* version (or service pack), you may see a few minor differences in the menu selections and options, which should not hinder you from learning the software.

KHC
Norman, Oklahoma
December 7, 2016

**About the Author**

Dr. Kuang-Hua Chang is a professor for the School of Aerospace and Mechanical Engineering at the University of Oklahoma (OU), Norman, OK. He received his diploma in Mechanical Engineering from the National Taipei Institute of Technology, Taiwan, in 1980; and M.S. and Ph.D. degrees in Mechanical Engineering from the University of Iowa in 1987 and 1990, respectively. Since then, he joined the Center for Computer-Aided Design (CCAD) at Iowa as a Research Scientist and CAE Technical Area Manager. In 1997, he joined OU. He teaches mechanical design and manufacturing, in addition to conducting research in computer-aided modeling and simulation for design and manufacturing of mechanical systems.

His work has been published in 9 books and more than 150 articles in international journals and conference proceedings. He has also served as a technical consultant to US industry and foreign companies, including LG-Electronics, Seagate Technology, etc. He is Associate Editor for two international journals, *Mechanics Based Design of Structures and Machines*, and *Computer-Aided Design and Applications*. He also serves on the Editorial Boards of *ISRN Mechanical Engineering*, *International Journal of Scientific Computing*, and *Journal of Software Engineering and Applications*.

**About the Cover Page**

The picture shown on the cover page is the solid model of a Formula SAE (Society of Automotive Engineers) style racecar designed and built by engineering students at the University of Oklahoma (OU) during 2005-2006. The racecar model was built in *Creo* with about 1,400 parts and assemblies. Even though this was a team effort, most parts and assemblies were created and managed by then-Senior Mechanical Engineering student, Mr. Dave Oubre (Super Dave). His dedication in creating such a detailed and accurate racecar solid model is admirable. His effort is highly appreciated.

Each year engineering students throughout the world design and build formula-style racecars and participate in the annual Formula SAE competitions. The result is a great experience for young engineers working on a meaningful engineering project as well as the opportunity to work in a dedicated team effort. The OU team has been competitive in the Formula SAE competitions. The team won numerous awards throughout the years, and finished 12th and 8th overall at the Formula SAE and Formula SAE West competitions, respectively, in 2006. Their 2005 racecar design also won the prestigious 2005 PTC Award in the Education, Colleges, and Universities category. The worldwide competition is sponsored by Parametric Technology Corporation.

A quarter of the racecar suspension was imported into *SOLIDWORKS* and was employed as an application example to be discussed in *Lesson 10* of this book. You will find more technical details of the racecar suspension in that lesson.

# Table of Contents

## Lesson 6: A Slider-Crank Mechanism

## Lesson 7: A Rail Carriage Example

## Lesson 8: A Compound Spur Gear Train

## Lesson 9: Cam and Follower

## Lesson 10: Kinematic Analysis of a Racecar Suspension

## Appendix A: Defining Joints

# Lesson 1: Introduction to *SOLIDWORKS Motion*

## 1.1 Overview of the Lesson

This lesson intends to provide you with a brief overview on *SOLIDWORKS Motion. SOLIDWORKS Motion*, formerly called *COSMOSMotion* (*SOLIDWORKS* 2008 and before), is a virtual prototyping tool that supports animation, analysis and design of mechanisms. Instead of building and testing physical prototypes of the mechanism, you may use *SOLIDWORKS Motion* (also called *Motion* in this book) to evaluate and refine the mechanism before finalizing the design and entering the functional prototyping stage. *Motion* will help you analyze and eventually design better engineering products that involve moving parts. More specifically, the software enables you to size motors and actuators, determine power consumption, layout linkages, develop cams, understand gear drives, size springs and dampers, and determine how contacting parts behave, which would usually require tests of physical prototypes. With such information, you will gain insight on how the mechanism works and why it behaves in certain ways. You will be able to modify the design and often achieve better design alternatives using the more convenient and less expensive virtual prototypes. In the long run, using virtual prototyping tools, such as *Motion*, will help you become a more experienced and competent design engineer.

In this lesson, we will start with a brief introduction to *Motion* and discuss various types of physical problems that *Motion* is capable of solving. We will then discuss capabilities offered by *Motion* for creating motion models, conducting motion analyses, and viewing motion analysis results. In the final section, we will preview examples employed in this book and topics to learn from these examples.

Note that materials presented in this lesson will be kept brief. More details on various aspects of the mechanism design and analysis using *Motion* will be given in later lessons.

## 1.2 What is *SOLIDWORKS Motion*?

*SOLIDWORKS Motion* is a computer software tool that supports engineers to analyze and design mechanisms. It is a module of the *SOLIDWORKS* product family developed by *Dassault Systèmes SOLIDWORKS Corporation*. This software supports users to create virtual mechanisms that answer general questions in product design as described next. A single piston engine shown in Figures 1-1 and 1-2 will be used to illustrate some typical questions, such as the following:

1.  Will the components of the mechanism collide or interfere in operation? For example, will the connecting rod collide with the inner surface of the piston or the inner surface of the engine case during operation?

2.  Will the components in the mechanism you design move according to your intent? For example, will the piston stay entirely in the piston sleeve? Will the system lock up when the firing force aligns vertically with the connecting rod?

3.   How much torque or force does it take to drive the mechanism? For example, what will be the minimum firing load to move the piston? Note that in this case, proper friction forces must be added to simulate the resistance between parts in contact in the mechanism before a realistic firing force can be calculated.

4.   How fast will the components move; e.g., the linear velocity of the piston?

5.   What is the reaction force or torque generated at a connection (also called *joint* or *constraint*) between components (or bodies) during motion? For example, what is the reaction force at the joint between the connecting rod and the piston pin? This reaction force is critical since the structural integrity of the piston pin and the connecting rod must be ensured; i.e., they must be strong and durable enough to sustain the firing load in operation.

The modeling and analysis capabilities in *Motion* will help you answer these common questions accurately and realistically, as long as the motion model is properly defined.

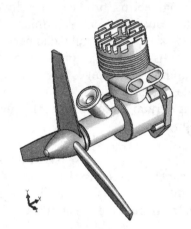

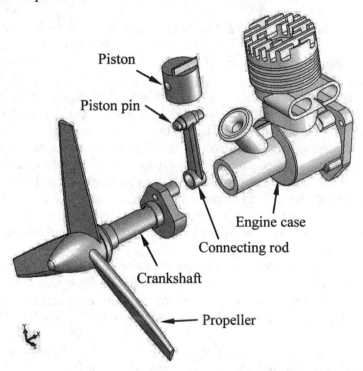

Figure 1-1  A Single Piston Engine
(Unexplode View)

Figure 1-2  Single Piston Engine (Explode View)

The capabilities available in *Motion* also help you search for better design alternatives. A better design alternative is very much problem dependent. It is critical that a design objective be clearly defined by the designer up front before searching for better design alternatives. For the engine example, a better design alternative could be a design that reveals:

1.   A smaller reaction force applied to the connecting rod, and
2.   No collisions or interference between components.

One of the common approaches for searching for design alternatives is to vary the component sizes of the mechanism. In order to vary component sizes for exploring better design alternatives, the parts and assembly must be adequately parameterized to capture design intents. At the parts level, design parameterization implies creating solid features and relating dimensions so that when a dimension value is changed, the part can be rebuilt properly. At the assembly level, design parameterization involves defining assembly mates and relating dimensions across parts. When an assembly is fully parameterized, a change in dimension value can be propagated automatically to all parts affected. Parts affected must be rebuilt successfully, and at the same time they will have to maintain proper position and orientation with respect to one another without violating any assembly mates or revealing part penetration or excessive gaps. For example, in this single-piston engine assembly, a change in the bore diameter of the engine case will alter not only the geometry of the engine case, but also all other parts affected, such as the piston, piston sleeve, and even the crankshaft, as illustrated in Figure 1-3. Moreover, they all have to be rebuilt properly and the entire assembly must stay intact through assembly mates.

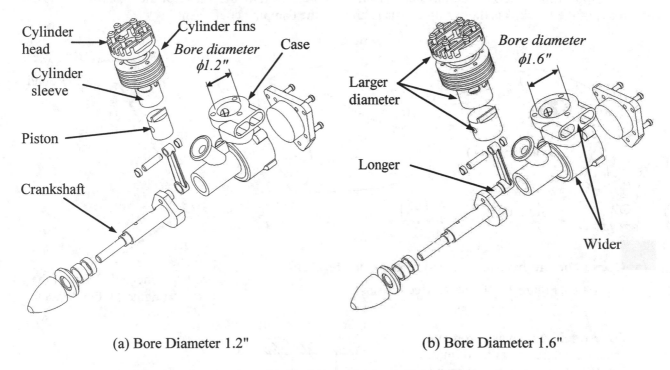

(a) Bore Diameter 1.2"                                        (b) Bore Diameter 1.6"

Figure 1-3  A Single-Piston Engine—Explode View

## 1.3  Mechanism Design and Motion Analysis

A mechanism is a mechanical device that transfers motion and/or force from a source to an output. It can be an abstraction (simplified model) of a mechanical system. A linkage consists of links (also called bodies), which are connected by joints (or connections), such as a revolute joint, to form open or closed chains (or loops, see Figure 1-4). Such kinematic chains, with at least one link fixed, become mechanisms. In this book, all links are assumed rigid. In general, a mechanism can be represented by its corresponding schematic drawing for discussion and presentation purposes. For example, a slider-crank mechanism characterizes the engine motion, as shown in Figure 1-5, which is a closed loop mechanism.

In general, there are two basic types of motion problems that you will have to solve in order to answer questions regarding mechanism analysis and design: kinematic and dynamic.

Kinematics is the study of motion without regard for the forces that cause the motion. A kinematic mechanism can be driven by a servomotor so that the position, velocity, and acceleration of each link of the mechanism can be analyzed at any given time. Typically, a kinematic analysis is conducted before the dynamic behavior of the mechanism can be simulated properly.

Dynamic analysis is the study of motion in response to externally applied loads. The dynamic behavior of a mechanism is governed by Newton's laws of motion. The simplest dynamic problem is the particle dynamics introduced in sophomore Dynamics class—for example, a spring-mass-damper system shown in Figure 1-6. In this case, motion of the mass is governed by the following equation derived from Newton's second law,

$$\sum F = p(t) - kx - c\dot{x} = m\ddot{x} \tag{1.1}$$

where (·) appearing on top of the physical quantities represents time derivative of the quantities, $m$ is the total mass of the block, $k$ is the spring constant, and $c$ is the damping coefficient.

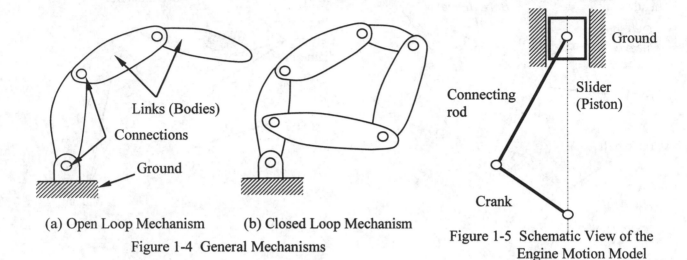

(a) Open Loop Mechanism        (b) Closed Loop Mechanism

Figure 1-4  General Mechanisms

Figure 1-5  Schematic View of the Engine Motion Model

For a rigid body, mass properties (such as the total mass, center of mass, moment of inertia, etc.) are taken into account for dynamic analysis. For example, motion of a pendulum shown in Figure 1-7 is governed by the following equation of motion,

$$\sum M = -mg\ell \sin\theta = I\ddot{\theta} = m\ell^2\ddot{\theta} \tag{1.2}$$

where $M$ is the external moment (or torque), $I$ is the mass moment of inertia of the pendulum, $m$ is the pendulum mass, $g$ is the gravitational acceleration, and $\ddot{\theta}$ is the angular acceleration of the pendulum.

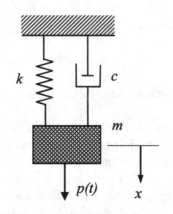

Figure 1-6  The Spring-Mass-Damper System

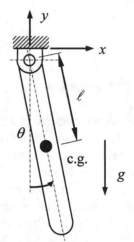

Figure 1-7  A Simple Pendulum

Dynamic analysis of a rigid body system, such as the single piston engine shown in Figure 1-3, is a lot more complicated than the single body problems. Usually, a system of differential and algebraic equations governs the motion and the dynamic behavior of the system. Newton's law must be obeyed by individual bodies in the system at all time. The motion of the system will be determined by the loads acting on the bodies or joint axes (e.g., a torque driving the system). Reaction forces and moments at the joints hold the bodies together.

Note that in *Motion*, you may create a kinematic analysis model; e.g., using a motion driver (linear or rotary motor) to drive the mechanism; and then carry out dynamic analyses. In dynamic analysis, position, velocity, and acceleration results are identical to those of kinematic analysis. However, the inertia of the bodies will be taken into account for analysis; therefore, reaction forces will be calculated between bodies.

## 1.4 *SOLIDWORKS Motion* Capabilities

### *Overall Process*

The overall process of using *SOLIDWORKS Motion* for analyzing a mechanism consists of three main steps: model generation, analysis (or simulation), and result visualization (or post-processing), as illustrated in Figure 1-8. Key entities that constitute a motion model include servo motors that drive the mechanism for kinematic analysis, external loads (force and torque), force entities such as spring and damper, and the initial conditions of the mechanism. Most importantly, assembly mates must be properly defined for the mechanism so that the motion model captures essential characteristics and closely resembles the behavior of the physical mechanism.

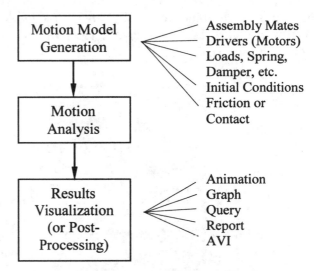

Figure 1-8  General Process of Using *SOLIDWORKS Motion*

The analysis or simulation capabilities in *Motion* employ simulation engine *ADAMS/Solver*, which solves the equations of motion for the mechanism. *ADAMS/Solver* calculates the position, velocity, acceleration, and reaction forces acting on each moving part in the mechanism. Typical simulation problems, including static (equilibrium configuration) and motion (kinematic and dynamic) are supported. More details about the analysis capabilities in *Motion* will be discussed later in this lesson.

The analysis results can be visualized in various forms. You may animate motion of the mechanism or generate graphs for more specific information, such as the reaction force of a joint in time domain. You may also query results at specific locations for a given time. Furthermore, you may ask for a report on results that you specified, such as the acceleration of a moving part in the time domain. You may also save the motion animation to an AVI for faster viewing and/or showing motion animation on other computers.

### *Operation Mode*

*Motion* is embedded in *SOLIDWORKS*. It is indeed an add-in module of *SOLIDWORKS*. Transition from *SOLIDWORKS* to *Motion* is seamless. All the solid parts, materials, assembly mates, etc. defined in *SOLIDWORKS* are automatically carried over into *Motion*. *Motion* can be accessed through menus and windows inside *SOLIDWORKS*. The same assembly created in *SOLIDWORKS* is directly employed for creating motion models in *Motion*. In addition, part geometry is essential for mass property computations in motion analysis. In *Motion*, all mass properties calculated in *SOLIDWORKS* are ready for use. In addition, the detailed part geometry supports interference check (as well as detects contact between bodies) for the mechanism during motion simulation in *Motion*.

### User Interfaces—MotionManager Window

User interface of the *Motion* is embedded in *SOLIDWORKS*. *SOLIDWORKS* users should find that it is straightforward to maneuver in *Motion*. The main interface for using *Motion* is through the *MotionManager* window, as shown in Figure 1-9. The *MotionManager* is a separate window that is used to create and play animations as well as conduct motion analysis. When you open an existing assembly (or part) in *SOLIDWORKS*, the *Motion Study* tab (with a default name *Motion Study 1*) will appear at the bottom of the graphics area. Clicking the *Motion Study* tab will bring up the *MotionManager* window. As shown in Figure 1-9, the user interface window of *MotionManager* consists of *MotionManager* tree (or *Motion* browser), *Motion* toolbar, filters, timeline area, etc.

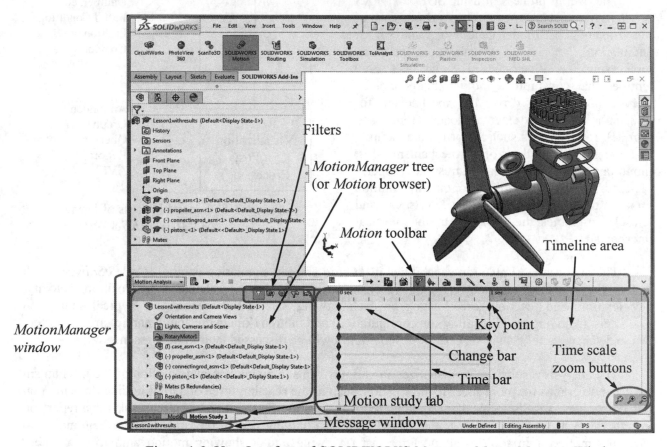

Figure 1-9  User Interface of *SOLIDWORKS Motion*—*MotionManager* Window

### MotionManager Tree (or Motion Browser)

Components are mapped from *SOLIDWORKS* assembly into *MotionManager* tree automatically, including root assembly, parts, sub-assemblies, and mates. An *Orientation and Camera Views* entity is also added. This entity will help you create precise animation in a specific view as desired, but has less to do with the motion simulation. Each part and sub-assembly entity can be expanded to show its composing components. Motion entities, such as spring and force, will be added to the tree once they are created. A result node will be added to the tree once a motion analysis is completed and result graphs are created. Similar to *SOLIDWORKS*, right clicking on a node in the *MotionManager* tree will bring up command options that you can choose to modify or adjust the *Motion* entity.

*Motion Toolbar*

The *Motion* toolbar shown in Figure 1-9 (and more detailed in Figure 1-10) provides major functions required to create and modify motion models, create and run analyses, and visualize results. The toolbar includes a type of study selection (for choosing from animation, basic motion, or motion analysis), calculate, animation controls, playback speed and mode options, save options, animation wizard, key point controls, simulation elements, results and graphs, and motion study property. As you move the mouse pointer over a button, a brief description about the functionality of the button appears. For example, if you move the mouse pointer close to the *Spring* button, a brief description about this button will appear next to it as well as in the *Message* window, located at the bottom left corner, as shown in Figure 1-9. Click some of the buttons and try to become more familiar with their functions. The buttons in *Motion* toolbar and their functions are also summarized in Table 1-1 with more details.

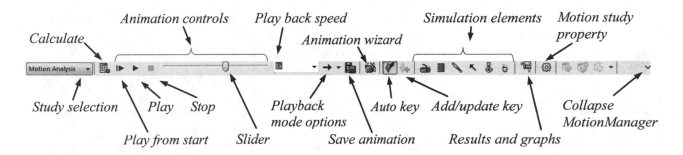

Figure 1-10  The *Motion* Toolbar

Note that you may choose *Animation*, *Basic Motion*, or *Motion Analysis* for study. You can use *Animation* to animate simple operation of assemblies, such as rotation, zoom in/out, explode/collapse assembly, etc. You may also add motors to animate simple kinematic motion of the assembly.

*Basic Motion* supports you for approximating the effects of motors, springs, collisions, and gravity on assemblies. *Basic Motion* takes mass into account in calculating motion. The computation is relatively fast, so you can use this for creating presentation-worthy animations using physics-based simulations.

*Motion Analysis* (offered by *SOLIDWORKS Motion*) provides you accurate simulation and analysis on the effects of motion elements (including forces, springs, dampers, and friction) on an assembly. *Motion Analysis* uses computationally strong kinematic solvers, and accounts for material properties as well as mass and inertia in the computations. You can also use *Motion Analysis* to plot simulation results for further analysis.

As a general rule of thumb, you may use *Animation* to create presentation-worthy animations for motion that does not require accounting for mass or gravity; use *Basic Motion* to create presentation-worthy approximate simulations of motion that account for mass, collisions, or gravity; and use *Motion Analysis* to run computationally strong simulations that take the physics of the assembly motion into consideration. *Motion Analysis* is the most computationally intensive of the three options.

Both *Animation* and *Basic Motion* are available in core *SOLIDWORKS*. *Motion Analysis* is only available with the *SOLIDWORKS Motion* add-in to *SOLIDWORKS*. Before using this book, you are encouraged to check with your system administrator to make sure the *Motion* module has been installed.

Table 1-1  The Shortcut Buttons in *Motion* Toolbar

| Symbol | Name | Function |
|---|---|---|
| | Calculate | Calculates the current simulation. If you alter the simulation, you must recalculate it before replaying. |
| | Play from start | Reset components and play the simulation. Use after simulation has been calculated. |
| | Play | Play the simulation beginning at the current timebar location. |
| | Stop | Stop the animation. |
| | Playback mode: Normal | Play from beginning to end once. |
| | Playback mode: Loop | Continuous play, from beginning to end, then loop to beginning and continue playing. |
| | Playback mode: Reciprocal | Continuous play, from beginning to end, then reverse—play from end to beginning. |
| | Save animation | Save the animation as an AVI movie file. |
| | Animation wizard | Allow to quickly create rotate model, explode or collapse animation. |
| | Auto key | Click to automatically place a new key when you move or change components. Click again to toggle this option. |
| | Add/update key | Click to add a new key or update properties of an existing key. |
| | Motor | Create a motor for motion analysis. |
| | Spring | Add a spring between two components. |
| | Damper | Add a damper between two components. |
| | Force | Create a force for motion analysis. |
| | Contact | Create a 3D contact between selected components. |
| | Gravity | Add gravity to the motion study. |
| | Results and graphs | Calculate results and create graphs. |
| | Motion study properties | Define motion study solution parameters. |
| | Collapse *MotionManager* | Collapse the *MotionManager* window. |

### *Filters*

Filters can be used to limit the *MotionManager* tree by including only animated, driving, selected or simulation results. Filters help make a cleaner display of entities in the *MotionManager* tree.

### *Timeline Area*

The area to the right of the *MotionManager* tree is the timeline area, which is the temporal interface for animation. The timeline area displays the times and types of animation events in the motion study. The timeline area is divided by vertical grid lines corresponding to numerical markers showing the time. The numerical markers start at 00:00:00. You may click and drag a key to define the beginning or end time of the animation or motion simulation.

After a simulation is completed, you will see several horizontal bars appear in the timeline area. They are *Change* bars in *Motion*. *Change* bars connect key points. They indicate a change between key points, which characterize the duration of animation, view orientation, etc. More about the timeline area will be discussed in *Lesson 2*.

Switching back and forth between *Motion* and *SOLIDWORKS* is straightforward. All you have to do is to click the *Model* tab (back to *SOLIDWORKS*) and *Motion Study* tab (to *Motion*) at the bottom of the graphics area.

### Defining Motion Entities

The basic entities of a valid *Motion* simulation model consist of ground parts (or ground body), moving parts (or moving bodies), constraints (imposed by mates in *SOLIDWORKS* assembly), initial conditions (usually position and velocity of a moving body), and forces and/or drivers. Each of the basic entities will be briefly discussed next. More details can be found in later lessons.

#### Ground Parts (or Ground Body)

A ground part, or a ground body, represents a fixed reference in space. The first component brought into the assembly is usually stationary, therefore becoming a ground part. Parts (or sub-assemblies) assembled to the stationary components without any possibility to move become part of the ground body. A symbol *(f)* is placed in front of the stationary components in the browser. You will have to identify moving and non-moving parts in your assembly, and use assembly mates to properly constrain the movement of the moving parts and completely fix the motion of non-moving parts.

#### Moving Parts (or Bodies)

A moving part or body is an entity representing a single rigid component (or link) that moves relatively to other parts (or bodies). A moving part may consist of a single *SOLIDWORKS* part or a sub-assembly composed of multiple parts. When a sub-assembly is designated as a moving part, none of its composing parts is allowed to move relative to one another within the sub-assembly.

A moving body has six degrees of freedom: three translational and three rotational (shown in Figure 1-11), while a ground body has none. That is, a moving body can translate and rotate along the *X*-, *Y*-, and *Z*-axes of a coordinate system. Rotation of a rigid body is measured by referring the orientation of its local coordinate system to the global coordinate system, which is shown at the lower left corner on the graphics area in *SOLIDWORKS*. In *Motion*, the local coordinate system is assigned automatically, usually at the mass center of the part or sub-assembly. Mass properties, including total mass, inertia, etc., are calculated using part geometry and material properties referring to the local coordinate system. A symbol *(-)* is placed in front of a moving component in the browser.

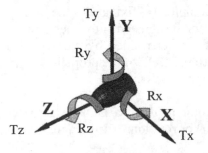

Tx: translational DOF along *X*-axis
Ty: translational DOF along *Y*-axis
Tz: translational DOF along *Z*-axis
Rx: rotational DOF along *X*-axis
Ry: rotational DOF along *Y*-axis
Rz: rotational DOF along *Z*-axis

Figure 1-11 Translational and Rotational Degrees of Freedom

*Constraints*

As mentioned earlier, an unconstrained rigid body in space has six degrees of freedom: three translational and three rotational. When you add a joint (or a constraint) between two rigid bodies, you remove degrees of freedom between them. Since *SOLIDWORKS* 2008, more commonly employed joints, such as revolute, translation, cylindrical joint, etc., in *Motion* have been replaced by assembly mates, such as coincident and concentric. Like joints, assembly mates remove degrees of freedom between parts.

Each independent movement permitted by a constraint (either a joint or an assembly mate) is a free degree of freedom. The free degrees of freedom that a constraint allows can be translational or rotational along the three perpendicular axes. For example, a concentric mate between the propeller assembly and the case of a single piston engine shown in Figure 1-12 allows one translational DOF (movement along the center axis, in this case, the *X*-axis of the global coordinate system, called reference triad in *SOLIDWORKS*) and one rotational DOF (rotating along *X*-axis). Since the case assembly is stationary, serving as the ground body, the propeller assembly has two free DOFs. Adding a coincident mate between the two respective faces of the engine case and the propeller shown in Figure 1-12 removes the remaining translational DOF, yielding a desired assembly that resembles the physical situation; i.e., with only the rotational DOF (along the *X*-axis).

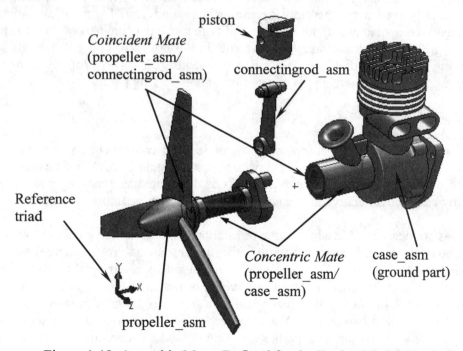

Figure 1-12  Assembly Mates Defined for the Engine Model (Explode View)

In creating a motion model, instead of completely fixing all the movements, certain DOFs (translational and/or rotational) are left to allow for designated movement. Such a movement will be either driven by a motor (for kinematic analysis) or determined by a force in dynamic analysis. For example, a rotary motor is created to drive the rotational DOF of the propeller in the engine example. This rotary motor rotates the propeller at a prescribed angular velocity.

In addition to velocity, you may use the motor to drive a DOF at a prescribed displacement or acceleration, either translation (using a linear motor) or rotational (using a rotary motor).

It is extremely important to understand assembly mates in order to create successful motion models. In addition to standard mates such as concentric and coincident (see Table 1-2 for a list of *SOLIDWORKS* standard mates), *SOLIDWORKS* provides advanced mates and mechanical mates (see Tables 1-3 and 1-4, respectively). Advanced mates provide additional ways to constrain or couple movements between bodies. For example, a linear coupler allows the motion of a translational or a rotational DOF of a given mate to be coupled to the motion of another mate. Also, the rotation motion of the propeller may be coupled to the translational motion of the piston (certainly if it makes sense). A coupler removes one additional degree of freedom from the motion model. Also, path mate (one of the advanced mates in *SOLIDWORKS*) allows a part to move along a curve slot, a groove, or fluting, varying its moving direction specified by the path curve. Such a capability supports animation and motion analysis of a different set of applications involving curvilinear motion. We will learn path mate in *Lesson 7*.

Table 1-2  Standard Mates of *SOLIDWORKS* Assembly

| Standard Mates | Descriptions |
|---|---|
| Coincident | Positions selected faces, edges, and planes (in combination with each other or combined with a single vertex) so they share the same infinite plane. Positions two vertices so they touch. |
| Parallel | Places the selected items so they remain a constant distance apart from each other. |
| Perpendicular | Places the selected items at a 90° angle to each other. |
| Tangent | Places the selected items tangent to each other (at least one selection must be a cylindrical, conical, or spherical face). |
| Concentric | Places the selections so that they share the same center line. |
| Lock | Maintains the position and orientation between two components. |
| Distance | Places the selected items with the specified distance between them. |
| Angle | Places the selected items at the specified angle to each other. |
| Default | Place the first part to the default coordinate system in assembly. |

Table 1-3  Advanced Mates of *SOLIDWORKS* Assembly

| Advanced Mates | Descriptions |
|---|---|
| Symmetric | Forces two similar entities to be symmetric about a plane or planar face. |
| Width | Centers a tab within the width of a groove. |
| Path | Constrains a selected point on a component to a path. |
| Linear/Linear Coupler | Establishes a relationship between the translation of one component and the translation of another component. |
| Limit | Allows components to move within a range of values for distance and angle mates. |

Table 1-4  Mechanical Mates of *SOLIDWORKS* Assembly

| Mechanical Mates | Descriptions |
|---|---|
| Cam | Forces a cylinder, plane, or point to be coincident or tangent to a series of tangent extruded faces. |
| Gear | Forces two components to rotate relative to one another about selected axes. |
| Hinge | Limits the movement between two components to one rotational degree of freedom. |
| Rack and Pinion | Linear translation of one part (the rack) causes circular rotation in another part (the pinion), and vice versa. |
| Screw | Constrains two components to be concentric, and also adds a pitch relationship between the rotation of one component and the translation of the other. |
| Universal Joint | The rotation of one component (the output shaft) about its axis is driven by the rotation of another component (the input shaft) about its axis. |

Mechanical mates include cam follower, gear, hinge, rack and pinion, screw, and universal joint. These are essential for motion model yet extremely easy to create in *Motion*. Some of these mates will be further discussed in later lessons.

In addition to mates, *Motion* provides contact constraint. The contact constraint helps to simulate physical problems involving contacts between bodies. Essentially, contact constraint applies a force to separate the parts when they are in contact and prevent them from penetrating each other. The contact constraint will become active as soon as the parts are touching (or close to be in contact). We will learn to create a contact constraint in *Lesson 3*.

## *Degrees of Freedom*

As mentioned earlier, an unconstrained body in space has six degrees of freedom. When assembly mates are added to assemble parts, constraints are imposed accordingly to restrict the relative motion between them.

Let us go back to the engine example shown in Figure 1-12. A concentric mate between the propeller and the engine case restricts movement on four DOFs (Ty, Tz, Ry, and Rz) so that only two movements are allowed, one translational (Tx) and one rotational (Rx). In order to restrict the translational movement, a coincident mate was added. A coincident mate between two respective faces of the propeller and the engine case restrict movement on three DOFs; i.e., Tx, Ry, and Rz. Even though combining these two mates gives desired rotational motion between the propeller and the case, there are redundant DOFs being restricted; i.e., Ry and Rz in this case.

It is important that you understand how to count the overall degrees of freedom for a motion model. For a given motion model, the number of degrees of freedom can be determined by using the Gruebler's count, defined as:

$$D = 6M - N - O \tag{1.3}$$

where $D$ is the Gruebler's count representing number of the free degrees of freedom of the mechanism, $M$ is the number of bodies excluding the ground body, $N$ is the number of DOFs restricted by all mates, and $O$ is the number of motion drivers (motors) defined in the system. For the motion model consisting of the propeller, the engine case, and the rotary motor, the Gruebler's count is:

$$D = 6 \times 1 - (4+3) - 1 = -2$$

However, we know that the propeller can only rotate along the *X*-axis; therefore, there is only one DOF for the system (Rx). The count should be *1*. After adding the rotary motor, the count becomes zero. The calculation gives us *–2* due to the fact that there are two redundant DOFs, Ry and Rz, which were restrained by both concentric and coincident mates. If we remove the redundant DOFs, the count becomes:

$$D = 6 \times 1 - (4+3-2) - 1 = 0$$

Another example, a door assembled to door frame using two hinge joints. Each hinge joint allows only one rotational movement along the axis of the hinge. The second hinge adds five redundant DOFs. The Gruebler's count becomes:

$$D = 6 \times 1 - 2 \times 5 = -4$$

Again, if we remove the redundant DOFs, the count becomes, as it should be:

$$D = 6 \times 1 - (2 \times 5 - 5) = 1$$

This is before any motor is added.

For kinematic analysis, the Gruebler's count must be equal to or less than *0* after adding motors. The *ADAMS/Solver* recognizes and deactivates redundant constraints during analysis. For a kinematic analysis, if you create a model and try to animate it with a Gruebler's count greater than *0*, the animation will not run and an error message will appear. For the door example, the vertical movement constrained by the second hinge is identified as redundant and removed from the solution. As a result, if conducting a dynamic simulation, the entire vertical force is carried by the first hinge. No (reaction) force will be calculated at the second hinge.

To get the Gruebler's count to zero, it is often required to replace mates that remove a larger number of constraints with mates that remove a smaller number of constraints and still restrict the mechanism motion in the same way. This is usually difficult if not entirely impossible. *Motion* detects the redundancies and ignores redundant DOFs in all analyses, except for dynamic analysis. In dynamic analysis, the redundancies lead to an outcome with a possibility of incorrect reaction results, yet the motion is correct. For complete and accurate reaction forces, it is critical that you eliminate redundancies from your mechanism if possible. The challenge is to find the mates that will impose non-redundant constraints and still allow for the intended motion. Very often this is not possible in *Motion* since instead of using regular joints (like revolute joint) it employs more primitive assembly mates for creating motion models. A concentric and a coincident mate in *SOLIDWORKS* is kinematically equivalent to a revolute joint, as illustrated in Figure 1-12 between the propeller and the engine case. A revolute joint removes five DOFs (with no redundancy); however, combining a concentric and a coincident mate removes seven DOFs, among which two are redundant. Using assembly mates to create motion models usually creates redundant DOFs.

The best strategy using *Motion* is to create an assembly that closely resembles the physical mechanism by using assembly mates that capture the characteristics of the motion revealed in the physical model. That is, first you should focus on creating an assembly that correctly captures the kinematic behavior of the mechanism. If you are conducting a dynamic analysis and want to capture reaction forces at critical components, you will have to examine the assembly mates and identify redundant DOFs. Then, check reaction forces at all mates for the relevant component and only take the reaction forces that make sense to you; mostly non-zero forces since zero reaction force is usually reported at the redundant DOFs by *Motion*.

*Forces*

Forces are used to operate a mechanism. Physically, forces are produced by motors, springs, dampers, gravity, etc. A force entity in *Motion* can be a force or torque. *Motion* provides three types of forces: applied forces, flexible connectors, and gravity.

Applied forces are forces that cause the mechanism to move in certain ways. Applied forces are very general, which involves defining the force direction and the force magnitude by specifying a constant force value or expression function, such as a harmonic function. The applied forces in *Motion* include action-only force or moment (where force or moment is applied at a point on a single rigid body, and no reaction forces are calculated), action and reaction force and moment, and impact force.

Flexible connectors resist motion and are simpler and easier to use than applied forces because you only supply constant coefficients for the forces, for instance a spring constant. The flexible connectors include translational springs, torsional springs, translational dampers, torsional dampers, and bushings.

A magnitude and a direction must be included for a force definition. You may select a predefined function, such as a harmonic function, to define the magnitude of the force or moment. For spring and damper, *Motion* automatically makes the force magnitude proportional to the distance or velocity between two points, based on the spring constant and damping coefficient entered, respectively. The direction of a force (or moment) can be defined by either along an axis defined by an edge or along the line between two points, where a spring or a damper is defined.

### *Initial Conditions*

In motion simulations, initial conditions consist of initial configuration of the mechanism and initial velocity of one or more components of the mechanism. Motion simulation must start with a properly assembled solid model that determines an initial configuration of the mechanism, composed by position and orientation of individual components. The initial configuration can be completely defined by assembly mates. However, one or more assembly mates will have to be suppressed, if the assembly is fully constrained, to provide adequate movement. In *SOLIDWORKS Motion*, initial velocity is defined as part of the definition of a moving part. The initial velocity can be translational or rotational along a direction spanned by the three axes.

### *Motion Drivers*

Motion drivers (or motors) are used to impose a particular movement on a free DOF over time. A motion driver specifies position, velocity, or acceleration as a function of time, and can control either translational or rotational motion. When properly defined, motion drivers will account for the remaining DOFs of the mechanism that bring the Gruebler's count to zero (exactly zero after removing all redundant DOFs) or less for a kinematic analysis.

### *Motion Simulation*

The *ADAMS/Solver* employed by *Motion* is capable of solving typical engineering problems, such as static (equilibrium configuration), kinematic, and dynamic, etc. Three numerical solvers are provided. They are GSTIFF, SI2_GSTIFF, and WSTIFF. GSTIFF is the default (numerical) integrator and is fast and accurate for displacements. It is used for wide range of motion simulations. SI2_GSTIFF provides better accuracy of velocities and accelerations but can be significantly slower. WSTIFF provides better accuracy for special problems, such as discontinuous forces.

Static analysis is used to find the rest position (equilibrium condition) of a mechanism, in which none of the bodies are moving. A simple example of the static analysis is illustrated in Figure 1-13, in which an equilibrium position of the block is to be determined according to its own mass $m$, the two spring constants $k_1$ and $k_2$, and the gravity $g$.

As discussed earlier, kinematics is the study of motion without regard for the forces or torque. A mechanism can be driven by a motion driver for a kinematic analysis, where the position, velocity, and acceleration of each link of the mechanism can be analyzed at any given time. Figure 1-14 shows a servomotor driving a slider-crank mechanism at a constant angular velocity.

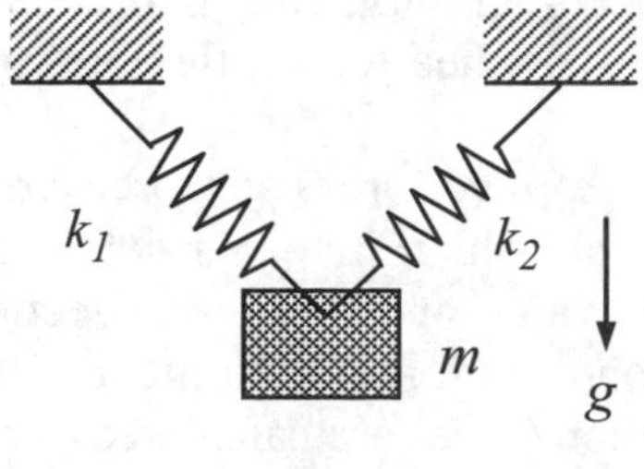

Figure 1-13 Static Analysis

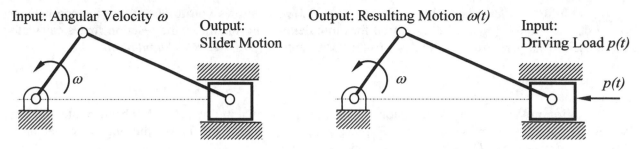

Figure 1-14  Kinematic Analysis                    Figure 1-15  Dynamic Analysis

Dynamic analysis is employed for studying the mechanism motion in response to loads, as illustrated in Figure 1-15, in which a force *p(t)* is applied at the slider in the horizontal direction to cause the slider to move, leading to the movement of the connection rod and the crank. This is the most complicated and common, and usually a more time-consuming analysis.

### *Viewing Results*

In *Motion*, results of the motion analysis can be visualized or presented using animations, graphs, reports, and queries. Animations show the configuration of the mechanism in consecutive time frames. Animations will give you a global view on how the mechanism behaves; for example, the single-piston engine shown in Figure 1-16. You may also export the animation to AVI for various purposes.

In addition, you may choose a joint (assembly mates in *Motion*) or a part to generate result graphs; for example, position vs. time of the mass center of the piston in the engine example shown in Figure 1-17. These graphs give you a quantitative understanding on the characteristics of the mechanism.

You may also query the results by moving the mouse pointer closer to the curve in a graph and leave the pointer for a short period. The result data will appear next to the pointer. In addition, you may ask *SOLIDWORKS Motion* for a report that includes a complete set of results output in the form of textual data or a Microsoft® Excel spreadsheet.

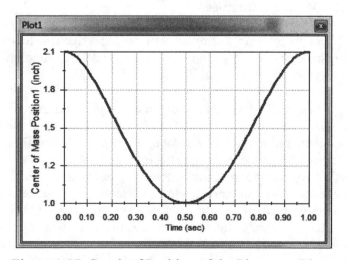

Figure 1-16  Motion Animation

Figure 1-17  Graph of Position of the Piston vs. Time

In addition to the capabilities discussed above, *Motion* allows you to check interference between bodies during motion (please see *Lesson 6* for more details). Furthermore, the reaction forces calculated can be used to support structural analysis using, for example, *SOLIDWORKS Simulation*.

## 1.5   Open Lesson 1 Model

A motion model for the single piston engine example shown in Figure 1-1 has been created for you. Download the files from www.sdcpublications.com, unzip them, and locate the engine model under *Lesson 1*. Copy or move *Lesson 1* folder to your hard drive.

Start *SOLIDWORKS* and open assembly *Lesson1withresults.SLDASM*. You should see an assembled engine model similar to that of Figure 1-1. Also, the *Motion Study* tab (with a default name *Motion Study 1*) will appear at the bottom of the graphics area. If you do not see this tab, you may not have *Motion* module installed properly. Consult with your IT staff to ensure a proper software install. Once the install is completed, you need to activate the *Motion* module by choosing from the pull-down menu

*Tools > Add-Ins*

In the *Add-Ins* window shown in Figure 1-18, click *SOLIDWORKS Motion* in both boxes (*Active Add-ins* and *Start Up*), and then click *OK*.

Click the *Motion Study* tab to bring up the *MotionManager* window. Note that a rotary motor, *RotaryMotor1*, has been added to the motion model. Click the *Play* button on the *Motion* toolbar; you should see that the propeller starts rotating, similar to that of Figure 1-16. Also, you may want to hide *case_asm* to see the motion inside the case, especially between the connection rod and the piston. You may do so by right clicking *case_asm<1>* from the *MotionManager* tree and choosing *Hide*.

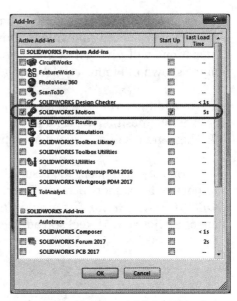

## 1.6   Motion Examples

More motion examples will be introduced in this book to illustrate the step-by-step details of modeling, simulation, and result visualization capabilities in *Motion*.

Figure 1-18   The *Add-Ins* Window

We will start with the same engine example in *Lesson 2*. *Lesson 2* will introduce numerous animations and basic motion study capabilities, with the focus on helping you become familiar with the basic operation in *MotionManager*, especially the timeline area. In *Lesson 3*, a simple ball-throwing example will be presented. This example will give you a quick run-through on using *Motion*. This lesson will also involve creating a contact constraint. *Lessons 4* through *9* focus on modeling and analysis of basic mechanisms and dynamic systems. In these lessons, you will learn assembly mates; forces and connections, including springs, gears, cam-followers; drivers and forces; various analyses; and graphs and results. *Lesson 10* illustrates the kinematic analysis of suspension of a ground vehicle. All examples and main topics to be discussed in each lesson are summarized in the following table.

Table 1-5  Examples Employed in This Book

| Lesson | Title | Motion Model | Problem Type | Things to Learn |
|---|---|---|---|---|
| 2 | Single Piston Engine Example | | Animation and basic motion | 1. This lesson focuses on helping you become familiar with the basic operations in *MotionManager*.<br>2. You will learn animation and basic motion study capabilities, especially using *Animation Wizard* to create simple animations in assembly. |
| 3 | Ball Throwing Example | | Particle Dynamics | 1. This lesson offers a quick run-through of general modeling and simulation capabilities in *Motion*.<br>2. You will learn the general process of using *Motion* to construct a motion model, run simulation, and visualize motion simulation results.<br>3. Gravity will be turned on and a contact constraint will be defined between the ball and the ground to simulate the ball bouncing scenario.<br>4. Simulation results are verified using analytical equations of motion. |
| 4 | Simple Pendulum | | Particle Dynamics | 1. This lesson provides a more in-depth discussion on using concentric and coincident mates to create a pendulum using *Motion*.<br>2. Simulation results are verified using analytical equations of motion. |
| 5 | Spring-Mass System | | Particle Dynamics | 1. This is a classical spring-mass system example you learned in Sophomore *Dynamics* class.<br>2. You will learn how to create a mechanical spring, align the block with the slope surface, and add an external force to pull the block.<br>3. Friction will be added between the block and the slope face.<br>4. Simulation results are verified using analytical equations learned in *Dynamics*. |

| 6 | Slider Crank Mechanism | | Multibody Dynamic Analyses | 1. This lesson uses a slider-crank mechanism to conduct kinematic and dynamic simulations.<br>2. You will learn to create motion driver as well as add a firing force for simulation.<br>3. The interference checking capability will be discussed.<br>4. Kinematic analysis results are verified using analytical equations of motion. |
|---|---|---|---|---|
| 7 | Rail-Carriage Example | | Path Mate | 1. This lesson introduces a useful and powerful mate, the path mate. Such a capability supports animation and motion analysis of a different set of applications involving curvilinear motion. |
| 8 | Compound Spur Gear Train | | Gear Train Analysis | 1. This lesson focuses on simulating motion of a spur gear train system.<br>2. You will learn how to use *SOLIDWORKS* and *Motion* to create a gear connection, and simulate the gear train motion.<br>3. Gear rotation speed is verified using analytical equations. |
| 9 | Cam and Follower | | Multibody Dynamic Analysis | 1. This lesson discusses cam and follower connection.<br>2. An inlet or outlet valve system of an internal combustion engine will be created and simulated.<br>3. Position and velocity of the valve will be graphed to monitor the motion of the system as well as assess the engineering design of the system. |
| 10 | Quarter suspension of a racecar | | Multibody Kinematic Analysis | 1. This lesson discusses kinematic analysis of a racecar suspension, including the addition of a cam representing the road profile.<br>2. Important kinematic measures of suspension design, including shock travel and camber angle, are included in discussion. |

# Lesson 2: Animations and Basic Motion —Single Piston Engine Example

## 2.1 Overview of the Lesson

In this lesson we focus on learning the basic operations in *MotionManager*. For those who are familiar with *SOLIDWORKS Animation* capabilities, this lesson may serve as a review. If you are not familiar with *Animation*, this lesson will help you learn *MotionManager*, especially in creating motion animation using *Animation* and *Basic Motion* studies. Skill you gained from this lesson should help you learn more in later lessons.

In this lesson, we will use the single piston engine discussed in *Lesson 1* as an example to illustrate the use of *MotionManager*. More specifically, we will use *Animation Wizard* to create animations for rotation, explode, and collapse of the engine assembly. Then, we will add a rotary motor to drive the propeller for a kinematic analysis. This is a very short and simple lesson. It should be an easy and fun lesson to go through.

## 2.2 The Single Piston Engine Example

### *Physical Model*

The engine example consists of four major components: engine case, propeller, connecting rod, and piston. The propeller is driven by a rotary motor at an angular speed of 60 rpm; i.e., one revolution per second. No gravity is present and English units system (IPS—inch, pound, second) is assumed.

### *SOLIDWORKS Parts and Assembly*

For this lesson, the parts and assembly have been created for you in *SOLIDWORKS*. There are eighteen parts and four assembly files. You can find these files at the publisher's web site (www.sdcpublications.com).

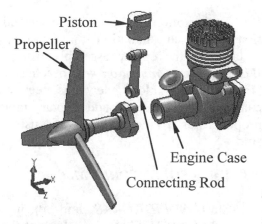

Figure 2-1  The Engine Example

We will start with *Lesson2.SLDASM*, in which the engine is properly assembled with one free degree of freedom. When the propeller is driven by the rotary motor, it will rotate, the crank shaft will drive the connecting rod, and the connection rod will push the piston up and down within the piston sleeve. Note that the assembly file *Lesson2withresults.SLDASM* contains a complete simulation model with simulation results. You may want to open this file to preview the motion model.

The assembly *Lesson2.SLDASM* consists of three sub-assemblies (*case_asm*, *propeller_asm*, and *connectingrod_asm*) and one part (*piston*). The *case_asm* is fixed (ground body). The *propeller_asm* is assembled to *case_asm* using concentric and coincident mates, as shown in Figure 2-2a. The propeller is free to rotate along the *X*-direction. The *connectingrod_asm* is assembled to the propeller (at the crankshaft) using concentric and coincident mates, as shown in Figure 2-2b. The connecting rod is free to rotate relative to the propeller (at the crankshaft) along the *X*-direction. Finally, the piston is assembled to the connecting rod (at pin) using a concentric mate, as shown in Figure 2-2c. The piston is also assembled to the case using another concentric mate. This mate restricts the piston movement along the *Y*-direction, which in turn restricts the top end of the connection rod to move vertically.

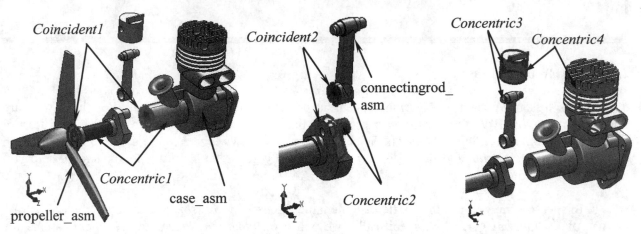

(a) Mates between Case and Propeller  (b) Mates between Propeller and Rod  (c) Mates between Piston and Rod, and Piston and Case

Figure 2-2  Assembly Mates Defined for the Engine Example

### *Motion Studies*

Two motion studies will be carried out. We will first create an animation for rotation, explode, and collapse of the engine assembly using the *Animation Wizard*, where *Animation* will be chosen for the study. Note that an explode view has been created for this example. Then, we will add a rotary motor to drive the propeller for a kinematic analysis using *Basic Motion* study.

### 2.3   Using *SOLIDWORKS Motion*

Start *SOLIDWORKS* and open assembly file *Lesson2.SLDASM*. First take a look at the explode view by selecting the root assembly (*Lesson2*), and press the right mouse button. In the menu appearing (Figure 2-3) choose *Explode*. You should see the assembly in an explode view similar to that of Figure 2-1. The animation will include exploding and collapsing the assembly. Right click the root assembly and choose *Collapse* to collapse the assembly.

Click the *Motion Study* tab (with the default name *Motion Study 1*) at the bottom of the graphics area to bring up the *MotionManager* window.

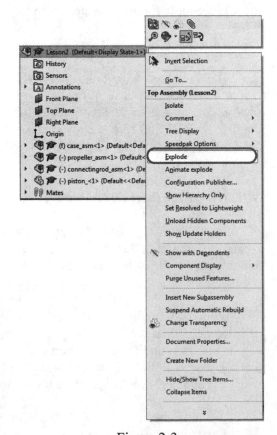

Figure 2-3

*Creating Animation*

We will create an animation that includes rotation, explode and collapse of the assembly using the *Animation Wizard*. *Animation Wizard* allows you to quickly create basic animations mentioned above.

<u>*Rotate Animation*</u>

Click the *Animation Wizard* button  from the *Motion* toolbar to bring up the *Animation Wizard* window (Figure 2-4a). Click *Rotate model* (default) and then *Next*.

The *Rotate model* option creates an animation by rotating the assembly around the *X*-, *Y*- or *Z*-axis. We will rotate the assembly along *Y*-axis. Select *Y-axis*, enter *1* for *Number of rotations*, and click *Clockwise* for the animation (Figure 2-4b). Click *Next*. Enter *Duration (seconds): 5* (see Figure 2-4c), and *Start Time (seconds): 0* (default), then click *Finish*.

Note that as soon as you click *Finish*, a five-second animation timeline appears in the timeline area, as shown in Figure 2-5. The *Orientation and Camera Views* feature receives key points to mark the changes in the orientations as the assembly rotates.

Choose *Animation* (default) for the study type (right above the *MotionManager* tree, as shown in Figure 2-5), and click *Play From Start* ▶ from the *Motion* toolbar to play the animation from the beginning. The engine assembly should make a complete turn along the *Y*-axis. Click the *Stop* button ■ and drag the slider (right next to the *Stop* button) back to the initial position.

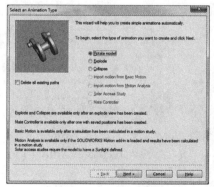
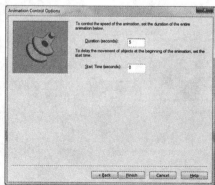

    (a) Select *Rotate model*            (b) Define Rotation Axis         (c) Define Animation Time Period

Figure 2-4  The Animation Wizards

<u>*Explode Animation*</u>

The *Explode* option creates an animation by converting the explode view information into an animation. The sequence and the distances of the individual explodes is used to create the movement. We will use the pre-defined explode view to create the animation.

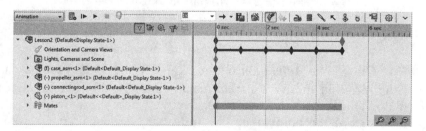

Figure 2-5  Timeline Area

Go back to *SOLIDWORKS* by clicking *Model* tab at the bottom of the graphics area. Select the *ConfigurationManager* above *SOLIDWORKS* browser. Expand the *Default* node, right click *ExpView1* and select *Explode*. You should see the explode view of the assembly appearing in the graphics area. Expand *ExpView1* to see the three explode steps, as shown in Figure 2-6. Click each step to see how they are defined.

Now, go back to *MotionManager* window (by clicking *Motion Study 1* at the bottom of the graphics area).

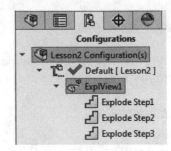

Figure 2-6 *ConfigurationManager*

Click the *Animation Wizard* button , choose *Explode* (Figure 2-7a) and then *Next*. Enter *Duration (seconds): 5*, (see Figure 2-7b) and *Start Time (seconds): 6* (that is, following rotation animation with one second gap), then click *Finish*. Note that as soon as you click *Finish*, the explode animation is added to the rotate animation with one second gap. An additional five-second animation timeline appears in the timeline area, as shown in Figure 2-8. Click *Play From Start* ▐► to play the animation. The assembly will rotate and then explode. Stop the animation and drag the slider back to the starting point.

*Collapse Animation*

The *Collapse* option creates an animation by reversing the *Explode* animation. Click the *Animation Wizard* again; choose *Collapse*. Enter *Duration (seconds): 5*, and *Start Time (seconds): 12* (following explode animation with one second gap). Another five-second animation timeline appears in the timeline area. Click *Play From Start* ▐► to play the animation. The assembly will rotate, explode and then collapse. Stop the animation and drag the slider back to the starting point.

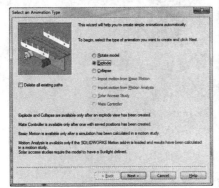
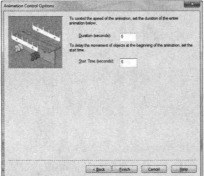
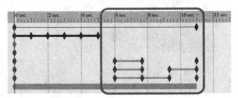

Figure 2-8  Added Timeline for Explode Animation

(a) Select *Explode*            (b) Define Animation Time Period

Figure 2-7  Animation Wizards

*Save as AVI*

Click *Save Animation* button  from the *Motion* toolbar, and use the default names and settings. Click *Save* in the *Save Animation to File* window (Figure 2-9). Click *OK* on the *Video Compression* dialog (Figure 2-10). In order to save the video, the animation will automatically play from the start and then record the animation. The time bar (vertical line) will run through the entire seventeen second animation period in the timeline area, and an *AVI* file (*Lesson2.avi*) will be created in the *Lesson 2* folder. You may use any *AVI* player, for example *Window Media Player*, to play the *AVI* video.

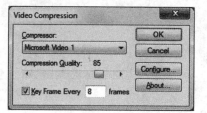

Figure 2-10

Figure 2-9

## *Rename the Motion Study*

Change the name of the motion study from *Motion Study 1* (default) to *Animation* by double clicking the *Motion Study* tab, and enter *Animation*. Save your model before moving to the next motion animation.

### *Creating a Motion Study using Basic Motion*

We will create the next motion study by adding a rotary motor to the propeller, and use *Basic Motion* to simulate the motion.

Create a new motion study by right clicking the *Motion Study* tab (renamed *Animation*), and choose *Create New Motion Study*, as shown in Figure 2-11. Rename the study *Basic Motion*. You should see a fresh timeline area.

Choose *Basic Motion* from the motion study selection (right above the *MotionManager* tree, see Figure 2-12). Click the *Motor* button from the *Motion* toolbar to bring up the *Motor* window (Figure 2-13). Choose *Rotary Motor* (default). Move the mouse pointer to the graphics area, and pick a circular arc of a part to define the rotational direction of the rotary motor; e.g., the circle on the drive washer of the propeller, as shown in Figure 2-14. A circular arrow appears indicating the rotational direction of the rotary motor.

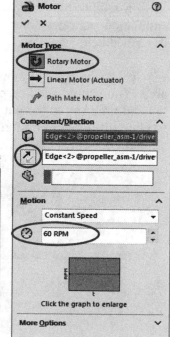

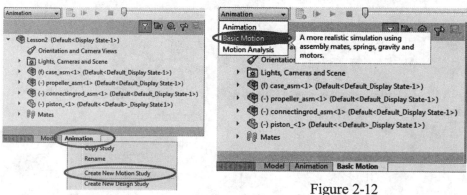

Figure 2-11

Figure 2-12

Figure 2-13

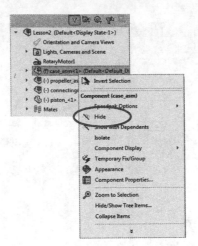

Figure 2-14 Pick an Arc                Figure 2-15 Hide Engine Case        Figure 2-16 Animation without Case

A counterclockwise direction is desired. You may change the direction by clicking the direction button ⟲ under *Component/Direction*. Choose *Constant speed* and enter *60* RPM for speed. Click the checkmark ✓ on top of the dialog box to accept the motor definition. You should see a *RotaryMotor1* added to the *MotionManager* tree.

Click *Calculate* button 🗔 from the *Motion* toolbar to simulate the motion. A default five second simulation will be carried out. The propeller should make five turns (recall we entered 60 RPM for the motor earlier), and a five-second animation timeline will be created in the timeline area.

You may want to hide the *case_asm* to see how the connecting rod and piston move. Right click *case_asm* in the *MotionManager* tree and select *Hide* (Figure 2-15). The case will be hidden from the graphics area. Play the animation again. You should now see the motion of connecting rod and piston like that of Figure 2-16.

Instead of repeating the same motion five times in five seconds, you may want to reduce the overall simulation period to one second and refine the time frame to create a smoother animation. Steps are described below.

Click the right key point (the end time key) of the change bar for the *Lesson2* node in the timeline area. You should see key point properties appear in the call out window, as shown in Figure 2-17. Drag the end time key to the one second mark. Note the changes of all bars in the timeline area. You may want to click the zoom in button (at the right lower corner) to expand the timeline area. Calculate and play the animation. Now the propeller should rotate only once per cycle.

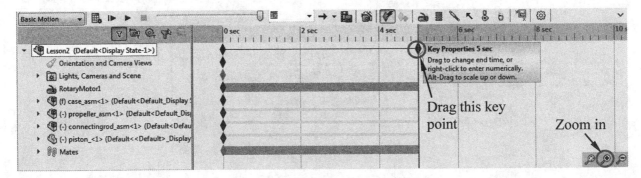

Figure 2-17

Click the *Motion Study Properties* button ⚙ from the *Motion* toolbar; the *Motion Study Properties* dialog appears (see Figure 2-18). Change *Frames per second* to *100* under *Basic Motion*, and then click the checkmark ✔ on top to accept the change. Calculate and animate the motion again.

Change the playback mode to loop ↺ and play the animation continuously. The animation looks great and it is fun to see objects moving, isn't it? Save the model and move to the next lesson.

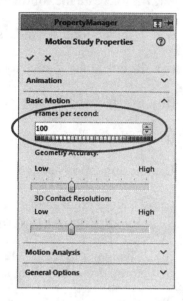

Figure 2-18 The *Motion Study Properties* Dialog Box

**Exercises:**

1. Open the propeller assembly and create an explode view like that of Figure E2-1. Use the *Animation Wizard* to create the following animations.

   (i)   Explode for 5 seconds, start time: 0 second.

   (ii)  Rotate along Y-axis for 5 seconds, start time: 6 second.

   (iii) Rotate along X-axis for 5 seconds, start time: 12 second.

   (iv)  Collapse for 5 seconds, start time: 18 second.

2. Use the animation created in Problem 1 with the following changes:

   (i)   Change crankshaft to red color for the entire animation. Hint: expand crankshaft node in the *MotionManager* tree, right click to choose *Appearance*, and select a red color.

   (ii)  Change spinner to blue color, and only activate the change for the first five seconds of the animation. Hint: drag key points.

   (iii) Right click the *Orientation and Camera Views* node in the *MotionManager* tree and select *Disable View Key Creation* (see Figure E2-2). Choose the *Front* view, and play the animation. What is the difference you see in the animation? Check *SOLIDWORKS Help* to find out more about the *Orientation and Camera Views*.

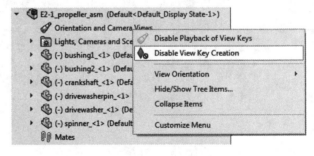

Figure E2-2

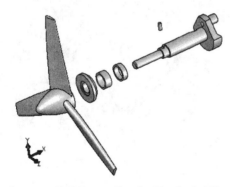

Figure E2-1  Propeller in Explode View

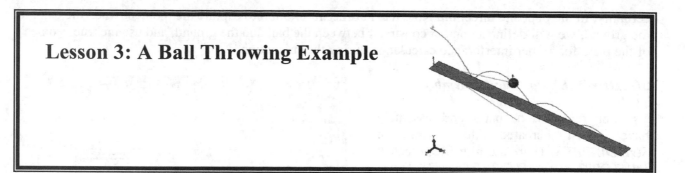

## 3.1   Overview of the Lesson

The purpose of this lesson is to provide you with a quick run-through on *SOLIDWORKS Motion*, using a simple ball-throwing example. This example simulates a ball thrown with an initial velocity at a given elevation. Due to gravity, the ball will travel following a parabolic trajectory and bounce back a few times when it hits the ground, as depicted in Figure 3-1. In this lesson, you will learn how to create a motion model to simulate the ball motion, run a simulation, and animate the ball motion. Simulation results obtained from *Motion* can be verified using particle dynamics theory that was learned in *Physics*. We will review the equations of motion, calculate the position and velocity of the ball, and compare our calculations with results obtained from *Motion*. Validating results obtained from computer simulations is extremely important. Note that *Motion* is not foolproof. In fact, a bug was found in *Motion 2017 SP1.0*, leading to erroneous results in this lesson due to a problem involved in the restitution coefficient. There is no doubt that it requires a certain level of experience and expertise to master the software. Before you arrive at that level, it will be indispensable for you to verify the simulation results whenever possible. Verifying the simulation results will increase your confidence in using the software and prevent you from being occasionally fooled by the erroneous simulations produced by the software. Note that often the erroneous results are due to modeling errors.

## 3.2   The Ball Throwing Example

***Physical Model***

The physical model of the ball example is very simple. The ball is made of *Cast Alloy Steel* with a radius of *10* in. The units system employed for this example is *IPS* (inch, pound, second). The gravitational acceleration is *386* in/sec². Note that you may check or change the units system by choosing from the pull-down menu

*Tools > Options*

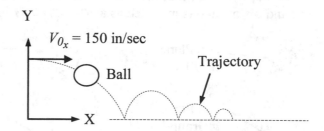

Figure 3-1   The Ball Throwing Example

and choose the *Document Properties* tab in the *Document Properties-Units* dialog box. Click the *Units* node, and then pick the units system you prefer, as shown in Figure 3-2. The *IPS* should have been selected.

The ball and ground are assumed rigid. The ball will bounce back when it hits the ground. A coefficient of restitution $C_R = 0.75$ is specified to determine the bounce velocity (therefore, the force) when the impact occurs. For this example, $C_R = V_f/V_i$, where $V_i$ and $V_f$ are the velocities of the ball before and after the impact. That is, the bounce velocity will be 75% of the incoming velocity in magnitude, and certainly, in the opposite direction. Note that in order for *Motion* to capture the moment when the ball hits the ground, we will define a contact constraint between the ball and the ground, and use the true geometry of the parts for a finer interference calculation during the simulation.

### SOLIDWORKS Parts and Assembly

For this lesson, parts and assembly have been created for you in *SOLIDWORKS*. There are four files created: *ball.SLDPRT*, *ground.SLDPRT*, *Lesson3.SLDASM*, and *Lesson3withresults.SLDASM*. You can find these files at the publisher's web site (www.sdcpublications.com). We will start with *Lesson3.SLDASM*, in which the ball is fully assembled to the ground; i.e., no movement is allowed. In addition, the assembly file *Lesson3withresults.SLDASM* consists of a complete simulation model with simulation results, in which some of the assembly mates were suppressed in order to provide adequate degrees of freedom for the ball to move. You may want to open this file to see how the ball is supposed to move.

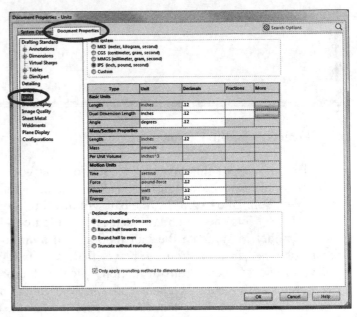

Figure 3-2  Checking Units System

Since the gravity is defined in the negative *Y*-direction of the global coordinate system, all parts and assembly are created for a motion simulation that complies with the gravity setting.

The assembly *Lesson3.SLDASM* consists of two parts: the ball (*ball.SLDPRT*) and the ground (*ground.SLDPRT*). The ball is fully assembled to the ground by three assembly mates of three pairs of reference planes. They are *Front* (ball)/*Front* (ground), *Right* (ball)/*Right* (ground), and *Top* (ball)/*Top* (ground), as shown in Figure 3-3. The distance between the reference planes *Top* (ball) and *Top* (ground) is *100* in., which defines the initial position of the ball, as shown in Figure 3-4. The radius of the ball is *10* in., and the ground is modeled as a *30×500×0.04* in.$^3$ rectangular block.

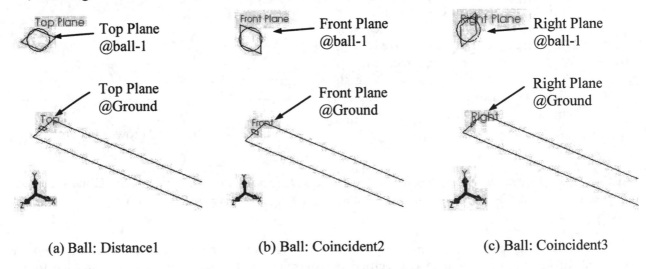

(a) Ball: Distance1                    (b) Ball: Coincident2                    (c) Ball: Coincident3

Figure 3-3  Assembly Mates Defined for the Example

Note that the *Y*-axis of the global coordinate system (or *Reference Triad* located at the lower left corner of the *SOLIDWORKS* graphics area, as shown in Figure 3-4) is pointing upward, which is consistent with the direction of the gravity but in the opposite direction.

### *Motion Model*

In this example, the ball will be the only movable body. Two assembly mates, *Distance1* and *Coincident3*, as shown in Figures 3-3a and 3-3c, will be suppressed to allow the ball to move on the *X-Y* plane. As mentioned earlier, the ball will be thrown with an initial velocity of $V_{0_x} = 150$ in/sec.

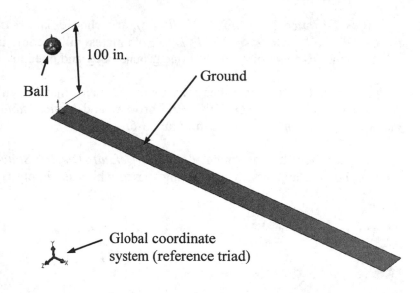

Figure 3-4  The *SOLIDWORKS* Assembly

A gravitational acceleration $-386$ in/sec$^2$ is defined in the *Y*-direction of the global coordinate system. The ball will reveal a parabolic trajectory due to gravity. A contact constraint will be added to characterize the impact between the ball and the ground. As discussed earlier, a coefficient of restitution $C_R = 0.75$ will be specified to determine the force that acts on the ball when the impact occurs. In this example, no friction is assumed.

### 3.3  Using *SOLIDWORKS Motion*

Start *SOLIDWORKS* and open assembly file *Lesson3.SLDASM*. Click the *Motion Study* tab at the bottom of the graphics area (*Motion Study 1*) to bring up the *MotionManager* window.

The *MotionManager* tree (or *Motion* browser) provides you with a graphical, hierarchical view of the motion model and allows you to edit existing motion entities conveniently by using the right click menu.

Switching back and forth between *Motion* and *SOLIDWORKS* assembly mode is straightforward. All you have to do is click the *Model* tab (to go back to *SOLIDWORKS*) and *Motion Study* tab (to *Motion*) when needed. As discussed in *Lesson 1*, the *Motion* toolbar, located at the top of the timeline area, provides settings, simulation and post-processing features for *Motion*. The *Play* button ▶ is especially handy when you finish running a simulation and are ready to animate the motion. Click some of the buttons and try to get familiar with their functions.

Before we start, we will suppress two assembly mates to allow the ball to move on the *X-Y* plane. Move the mouse pointer to *SOLIDWORKS* browser, and expand the *Mates* branch by clicking the small ⊞ button in front of it. You should see three mates constraints: *Distance1(Ground<1>,ball<1>)*, *Coincident2(Ground<1>,ball<1>)*, and *Coincident3(Ground<1>,ball<1>)* listed in the browser, as shown in Figure 3-5.

Pick *Distance1(Ground<1>,ball<1>)*, and choose *Suppress* 🔒 (see Figure 3-5). Repeat the same for *Coincident3(Ground<1>,ball<1>)*. Both mates will become inactive, and the ball is free to move on the *XY* plane. At this point, there is one ground body and one moving body (ball) in the motion model.

Please note that sometimes the suppressed mates may be still active. It is a good idea to expand the *Mates* node in both the *SOLIDWORKS* browser and *MotionManager* tree, and make sure the suppressed mates are grayed out, as shown in Figure 3-6.

If you click the root node, *Lesson3 (Default<Display State-1>)* in the browser, you should see a rectangular box appears in the graphics area, which is simply the bounding box for the assembly (see Figure 3-7).

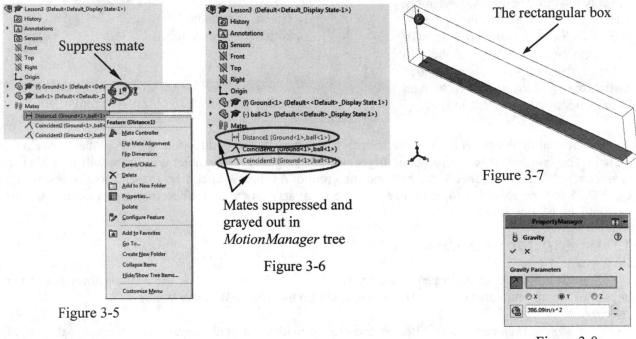

Figure 3-5

Figure 3-6

Figure 3-7

Figure 3-8

### *Defining Gravity*

Click the *Gravity* button 🔵 from the *Motion* toolbar to bring up the *Gravity* dialog box. Choose *Y* and keep the g-value (386.09in/s^2) in the dialog box, as shown in Figure 3-8. In the graphics area, an arrow appears at the right lower corner ⬇, pointing downward indicating the direction of the gravity.

Click the checkmark ✅ on top of the dialog box to accept the gravity. A gravity node (*Gravity*) should appear in the *MotionManager* tree.

### *Defining Initial Velocity*

We will add an initial velocity $V_{0_x} = 150$ in/sec to the ball. Click *ball* from the *MotionManager* tree, press the right mouse button, and choose *Initial Velocity* (see Figure 3-9a); the *Initial Velocity* dialog box will appear (Figure 3-9b).

From the graphics area, pick an edge on the ground part that is parallel to the *X*-axis like that in Figure 3-9c. Enter *150in/s*, as shown in Figure 3-9b. Click the checkmark ✅ to accept the initial velocity.

### Running Simulation

Choose *Motion Analysis* for the study (see Figure 3-10). Click the *Motion Study Properties* button ⚙️ from the *Motion* toolbar. In the *Motion Study Properties* dialog box (Figure 3-11), enter *100* for *Frames per second*, and click the checkmark on top. Drag the end time key to the three second mark (see Figure 3-12) in the timeline area to define the simulation duration. Click the *Zoom In* button to zoom in the timeline area if needed.

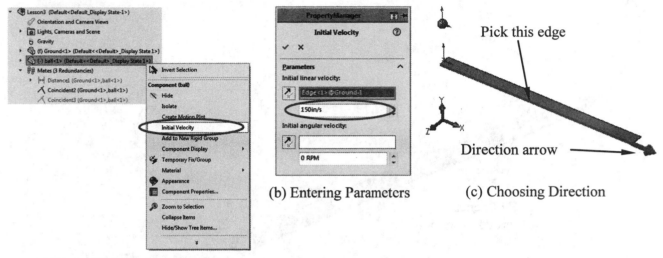

(a) Right Clicking the *ball* Node

(b) Entering Parameters

(c) Choosing Direction

Figure 3-9  Defining Initial Velocity

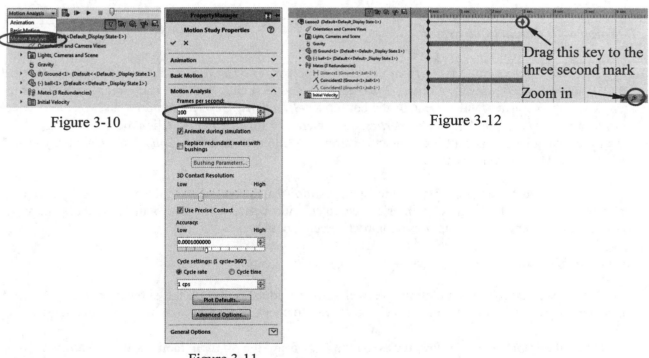

Figure 3-10

Figure 3-11

Figure 3-12

Click the *Calculate* button from the *Motion* toolbar to simulate the motion. A three second simulation will be carried out. The ball will be dropped from 100 in. above the ground with an initial horizontal velocity 150 in/sec. After a few seconds, the ball starts moving. The ball falls through the ground and continues falling, as shown in Figure 3-13, which is not realistic. Note that the trace path that indicates the trace of the center of the ball is turned on in Figure 3-13. We will learn how to do this later. For the time being we will add a contact constraint between the ball and the ground in order to make the ball bounce back when it hits the ground. The contact constraint will create a force to prevent the ball from penetrating the ground. This constraint will only be activated if the ball comes into contact with the ground.

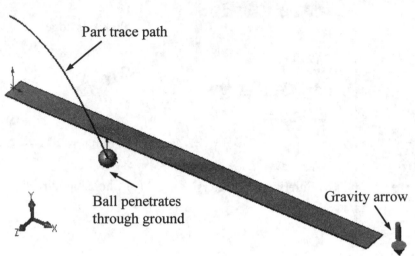

Figure 3-13   Animation: Ball Penetrating Through the Ground

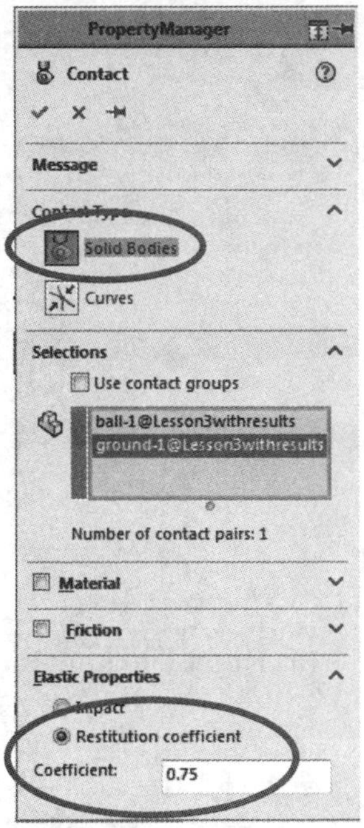

Figure 3-14

### *Defining a 3D Contact Constraint*

Click the *Contact* button from the *Motion* toolbar. The *Contact* dialog box will appear (Figure 3-14). Choose *Solid Bodies* for *Contact Type* (default), and then click both ball and ground in the graphics area. These two entities will appear in the box under *Selections*. Deselect *Material* and *Friction*, select *Restitution coefficient* for *Elastic Properties*, and enter *0.75* for *Coefficient*. Click the checkmark on top of the dialog box to accept the contact constraint. A contact constraint (*Solid Body Contact1*) should appear in the *MotionManager* tree.

At each time frame during the simulation, *Motion* calculates whether the parts interfere. If they interfere, *Motion* performs a finer interference calculation between the two bodies. At the same time, *ADAMS/Solver* computes and applies an impact force on both bodies.

### *Rerun the Simulation*

Before we rerun the simulation, we will have to adjust some of the simulation parameters. In particular, we will ask *Motion* to use precise geometry to check contact in each frame of simulation.

Click the *Motion Study Properties* button from the *Motion* toolbar. In the *Motion Study Properties* dialog box (Figure 3-15), choose *Use Precise Contact*. Click the checkmark to accept the contact constraint. Also, bring the ball to its initial position by dragging the time line (vertical line as shown in Figure 3-16) to 0 sec in the timeline area, or drag the slider right next to the *Stop* button in the *Motion* toolbar all the way to the left (beginning).

Click the *Calculate* button again to rerun the simulation. After a few seconds, the ball starts moving. As shown in Figure 3-17 (no trace path shown yet) the ball will hit the ground and bounce back a few times before the simulation ends. The ball did not fall through the ground this time due to the addition of the contact constraint. We will learn how to create a trace path next. Note that if you are using *Motion 2017 SP1.0*, you will see the trace of the ball like that of Figure 3-17, which is incorrect. Verification of the simulation results is included in Section 3.4.

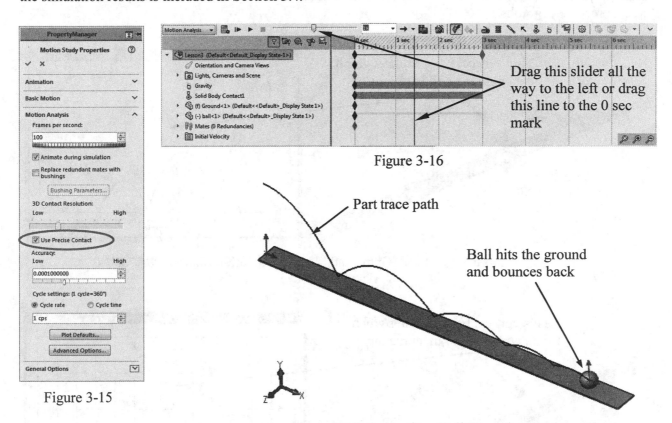

Figure 3-16

Figure 3-15

Part trace path

Ball hits the ground and bounces back

Figure 3-17  Animation: Ball Bouncing Back

### *Displaying Simulation Results*

*Motion* allows you to graphically display the path that any point on any moving part follows. This is called a trace path. Note that the trace path of the ball was displayed in both Figures 3-13 and 3-17. We will learn how to create a part trace path.

Click the *Results and Plots* button ![icon] from the *Motion* toolbar. In the *Results* dialog box (Figure 3-18), choose *Displacement/Velocity/Acceleration* from the first pull-down field. Then select *Trace Path* for the sub-category. Pick the origin point of the ball from the graphics area, and click the checkmark. A *Results* (and *Plot1<Trace Path1>*) node will be added to the *MotionManager* tree. Click the *Calculate* button again. When the ball starts moving, you should see a trace path following the ball like that of Figure 3-17.

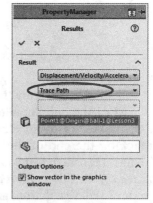

Figure 3-18

Next, we will create a graph for the *Y*-position of the ball. Right click *ball* in the *MotionManager* tree, and choose *Create Motion Plot*. In the *Results* dialog box (refer to Figure 3-18), choose *Displacement/Velocity/Acceleration*, select *Center of Mass Position*, and then *Y Component*. Click the checkmark.

An XY plot of the center of mass of the ball in the *Y*-direction will appear, similar to that of Figure 3-19 (which is incorrect) if you are running *Motion 2017 SP1.0*. Figure 3-20 shows correct results obtained from an earlier version of *Motion* (more precisely *COSMOSMotion* 2008). Note that you may adjust properties of the graph, for instance the axis scales, following steps similar to those of Microsoft® Excel spreadsheet graphs.

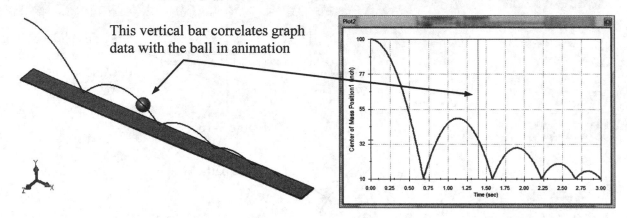

Figure 3-19  Result Graph and Motion Animation (*Motion 2017 SP1.0*)

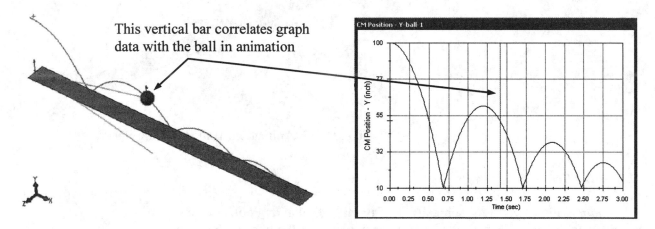

Figure 3-20  Correct Result Graph and Motion Animation (Recorded from an Earlier Version)

Both graphs show that the ball was thrown from *Y = 100* in. and hits the ground at *Y = 10* in. (which is CM of the ball since the ball radius is 10 in.) at a time about *t = 0.64* seconds. The ball bounces back and moves up to an elevation determined by the coefficient of restitution. The motion continues until reaching the end of the simulation. In the following discussion, we only refer to the correct simulation results recorded from an earlier version. You may open the assembly file *Lesson3withCorrectResults (saved from 2008 versions).SLDASM* under a subfolder *Lesson3withCorrectResults* of *Lesson 3* folder to review the simulation and result graphs (click *Play from Start* button ▶, do not click the *Calculate* button 🗐). Those who are running *2017 SP1.0* may see graphs with different results.

Note that you may click any location in the graph to bring up a fine red vertical line that correlates the graph with the position of the ball in animation. As shown in the graph of Figure 3-20, at *t* = *1.4* seconds, the ball is roughly at *Y* = *44* in. The snapshot of the ball at that specific time and its *Y*-location in the graph is shown in the graphics area.

Add two more graphs for the ball, the *X*-velocity and *Y*-velocity. As shown in Figure 3-21, the *X*-velocity of the ball is a constant *150* in/sec, which is equal to the initial velocity. This is because we turned off all friction for the contact constraint earlier (see Figure 3-14); therefore, no energy loss due to contact. Figure 3-22 shows *Y*-velocity of the ball. The ball moves with a linear velocity due to gravity. The *Y*-velocity is about *−263* in/sec at *t* = *0.67* seconds. You may see this data by moving the point close to the corner point of the curve and leaving the point for a short period. The position data will appear.

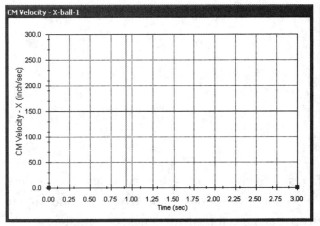

Figure 3-21  *X*-Velocity of the Ball CM

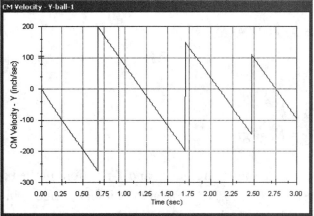

Figure 3-22  *Y*-Velocity of the Ball CM

You may also convert the *XY Plot* data to Microsoft® Excel spreadsheet by simply moving the pointer inside the graph and right click to choose *Export CSV*. Open the spreadsheet to see more detailed simulation data. From the spreadsheet (Figure 3-23), the *Y*-velocity right before and after the ball hits the ground is *−264* and *198* in/sec, respectively. The ratio of *198/264* is about *0.75*, which is the coefficient of restitution we defined earlier. Note that the spreadsheet results were obtained from an earlier version.

If you use *Motion 2017 SP1.0*, the results you obtained are incorrect. If you check the *Y*-velocity data before and after the ball hits the ground to calculate the restitution coefficient, it does not equal 0.75. In addition, the coefficient is not even a constant at the other bounce points, if you check more locations. This result is obviously incorrect.

You may save the animation as an *AVI* movie file, following steps described in *Lesson 2*.

Be sure to save the model before exiting from *Motion*.

## 3.4  Result Verifications

In this section, we will verify analysis results obtained from *Motion* using particle dynamics theory we learned in *Physics*.

| | A | B | C |
|---|---|---|---|
| 111 | 0.648 | -250.271 | |
| 112 | 0.654 | -252.588 | |
| 113 | 0.66 | -254.906 | |
| 114 | 0.666 | -257.223 | |
| 115 | 0.672 | -259.54 | |
| 116 | 0.678 | -261.857 | |
| 117 | 0.684 | -264.175 | |
| 118 | 0.684362 | 198.0581 | |
| 119 | 0.69 | 195.8807 | |
| 120 | 0.696 | 193.5633 | |
| 121 | 0.702 | 191.246 | |
| 122 | 0.708 | 188.9287 | |
| 123 | 0.714 | 186.6114 | |
| 124 | 0.72 | 184.2941 | |
| 125 | 0.726 | 181.9767 | |
| 126 | 0.732 | 179.6594 | |
| 127 | 0.738 | 177.3421 | |
| 128 | 0.744 | 175.0248 | |
| 129 | 0.75 | 172.7074 | |
| 130 | 0.756 | 170.3901 | |

CM Velocity - Y-ball-1.csv

Figure 3-23

There are two assumptions that we have to make in order to apply the particle dynamics theory to this ball-throwing problem:

(i) The ball is of a concentrated mass, and
(ii) No air friction is present.

### *Equation of Motion*

It is well-known that the equations of motion that describe the position and velocity of the ball are, respectively,

$$P_x = P_{0_x} + V_{0_x} t \tag{3.1a}$$

$$P_y = P_{0_y} + V_{0_y} t - \frac{1}{2} g t^2 \tag{3.1b}$$

and

$$V_x = V_{0_x} \tag{3.2a}$$

$$V_y = V_{0_y} - gt \tag{3.2b}$$

where $P_x$ and $P_y$ are the $X$- and $Y$-positions of the ball, respectively; $V_x$ and $V_y$ are the $X$- and $Y$-velocities, respectively; $P_{0x}$ and $P_{0y}$ are the initial positions in the $X$- and $Y$-directions, respectively; $V_{0x}$ and $V_{0y}$ are the initial velocities in the $X$- and $Y$-directions, respectively; and $g$ is the gravitational acceleration.

These equations can be implemented using, for example, Microsoft® Excel spreadsheet shown in Figure 3-24, for numerical solutions. In Figure 3-24, Columns B and C show the results of Eqs. 3.1a and 3.1b, respectively; with a time interval from *0* to *3* seconds and increment of *0.01* seconds. Columns D and E show the results of Eqs. 3.2a and 3.2b, respectively.

Note that when the ball hits the ground, we will have to reset the velocity to 75% of that at the prior time step and in the opposite direction.

Data in column C is graphed in Figure 3-25. Columns D and E are graphed in Figure 3-26. Comparing Figure 3-25 with Figure 3-20 and Figure 3-26 with Figures 3-21 and 3-22, the results obtained from theory and *Motion* should be very close (not the case for *Motion 2017 SP1.0*), which means the dynamic model has been created properly in *Motion*. Note that the solution spreadsheet can be found at the publisher's website (filename: *lesson3.xls*).

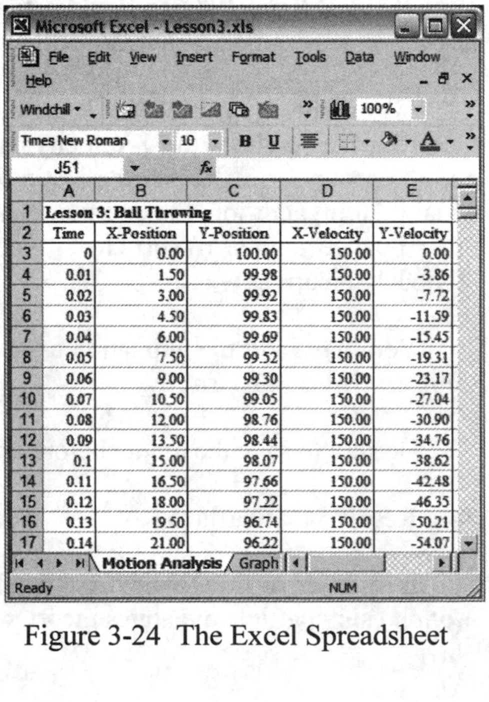


| | A | B | C | D | E |
|---|---|---|---|---|---|
| 1 | Lesson 3: Ball Throwing | | | | |
| 2 | Time | X-Position | Y-Position | X-Velocity | Y-Velocity |
| 3 | 0 | 0.00 | 100.00 | 150.00 | 0.00 |
| 4 | 0.01 | 1.50 | 99.98 | 150.00 | -3.86 |
| 5 | 0.02 | 3.00 | 99.92 | 150.00 | -7.72 |
| 6 | 0.03 | 4.50 | 99.83 | 150.00 | -11.59 |
| 7 | 0.04 | 6.00 | 99.69 | 150.00 | -15.45 |
| 8 | 0.05 | 7.50 | 99.52 | 150.00 | -19.31 |
| 9 | 0.06 | 9.00 | 99.30 | 150.00 | -23.17 |
| 10 | 0.07 | 10.50 | 99.05 | 150.00 | -27.04 |
| 11 | 0.08 | 12.00 | 98.76 | 150.00 | -30.90 |
| 12 | 0.09 | 13.50 | 98.44 | 150.00 | -34.76 |
| 13 | 0.1 | 15.00 | 98.07 | 150.00 | -38.62 |
| 14 | 0.11 | 16.50 | 97.66 | 150.00 | -42.48 |
| 15 | 0.12 | 18.00 | 97.22 | 150.00 | -46.35 |
| 16 | 0.13 | 19.50 | 96.74 | 150.00 | -50.21 |
| 17 | 0.14 | 21.00 | 96.22 | 150.00 | -54.07 |

Figure 3-24 The Excel Spreadsheet

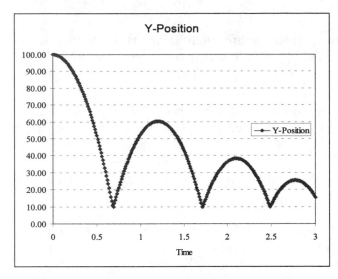

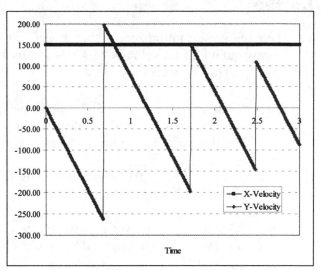

Figure 3-25  Graph of the *Y*-Position of the
            Ball Obtained from Spreadsheet
            Calculations

Figure 3-26  Graph of the *X*- and *Y*-Velocities of
            the Ball Obtained from Spreadsheet
            Calculations

**Exercises:**

1.  Use the same motion model to conduct a simulation for a different scenario. This time the ball is thrown at an initial velocity of $V_{0_x} = 100$ in/sec and $V_{0_y} = 50$ in/sec from an elevation of *150* in., as shown in Figure E3-1.

    (i)   Create a dynamic simulation model using *Motion* to simulate the trajectory of the ball. Report position, velocity, and acceleration of the ball at *0.2* seconds in both vertical and horizontal directions obtained from *Motion*.

    (ii)  Derive and solve the equations that describe the position and velocity of the ball. Compare your solutions with those obtained from *Motion*.

    (iii) Calculate the time for the ball to reach the ground and the distance it travels. Compare your calculation with the simulation results obtained from *Motion*.

2.  A *1"×1"×1"* block slides from top of a *45°* slope (due to gravity) without friction, as shown in Figure E3-2. The material of the block and the slope is *AL2014*.

    (i)   Create a dynamic simulation model using *Motion* to analyze motion of the block. Report position, velocity, and acceleration of the block in both vertical and horizontal directions at *0.5* seconds obtained from *Motion*.

    (ii)  Create a *LimitDistance* mate to stop the block when its front lower edge reaches the end of the slope. You may want to review *SOLIDWORKS* help menu or preview *Lesson 5* to learn more about the *LimitDistance* mate.

    (iii) Derive and solve the equation of motion for the system. Compare your solutions with those obtained from *Motion*.

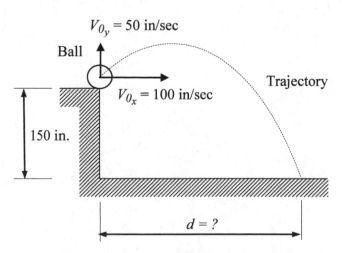

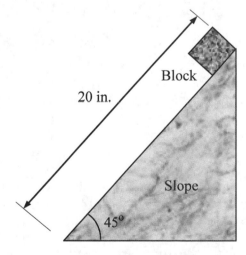

Figure E3-1  The Ball Throwing Problem                    Figure E3-2  The Block Sliding Problem

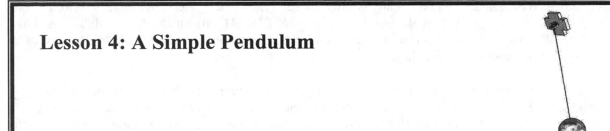

# Lesson 4: A Simple Pendulum

## 4.1 Overview of the Lesson

In this lesson, we will create a simple pendulum motion model using *SOLIDWORKS Motion*. The pendulum will be released from a position slightly off from the vertical line. Once released, the pendulum will rotate freely due to gravity. In this lesson, we will learn how to create a pendulum motion model, run a dynamic analysis, and visualize the analysis results. The dynamic analysis results of the simple pendulum example can be verified using particle dynamics theory. Similar to *Lesson 3*, we will formulate the equation of motion; calculate the angular position, velocity, and acceleration of the pendulum; and compare our calculations with results obtained from *Motion*.

## 4.2 The Simple Pendulum Example

### *Physical Model*

The physical model of the pendulum is composed of a sphere and a rod rigidly connected, as shown in Figure 4-1. The radius of the sphere is *10* mm. The length and radius of the thin rod are *90* mm and *0.5* mm, respectively. The top of the rod is connected to the wall with a revolute joint (assembled using concentrate and coincident mates). This revolute joint allows the pendulum to rotate. Both rod and sphere are made of Alunimum. Note that from the *SOLIDWORKS* material library the *Aluminum Alloy 2014* has been selected for both sphere and rod. The *MMGS* units system is chosen for this example (millimeter for length, Newton for force, and second for time). Note that in the *MMGS* units system, the gravitational acceleration is *9,806* mm/sec$^2$.

The pendulum will be released from an angular position of *10* degrees measured from the vertical position about the rotational axis of the revolute joint. The rotation angle is intentionally kept small so that the particle dynamics theory can be applied to verify the simulation result.

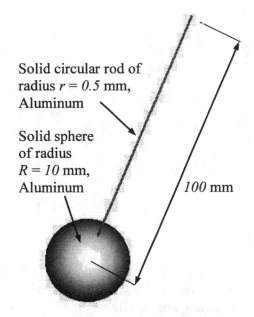

Solid circular rod of radius *r = 0.5* mm, Aluminum

Solid sphere of radius *R = 10* mm, Aluminum

*100* mm

Figure 4-1 The Pendulum Physical Model

### *SOLIDWORKS Parts and Assembly*

In this lesson, *SOLIDWORKS* parts of the pendulum examples have been created for you. There are four model files created: *pendulum.SLDPRT*, *ground.SLDPRT*, *Lesson4.SLDASM*, and

*Lesson4withresults.SLDASM.* You can find these files at the publisher's web site (www.sdcpublications.com). We will start with *Lesson4.SLDASM*, in which the pendulum is fully assembled to the ground. In addition, the assembly file *Lesson4withresults.SLDASM* consists of a complete simulation model with simulation results.

In the assembly models, there are three assembly mates, *Concentric1(ground<1>,pendulum<1>)*, *Coincident1(ground<1>, pendulum <1>)*, and *Angle1(pendulum <1>, Right Plane)*, as shown in Figure 4-2. Note that the first two mates restrict pendulum to rotate about the *Z*-axis of the global coordinate system. The third mate, *Angle1*, rotates the pendulum a negative *10* degree angle about the *Z*-axis. This mate is required to position the pendulum for an initial condition. You may want to use the *Front* view (one of *SOLIDWORKS* predefined views located on top of the graphics area) to see how the pendulum is oriented. The pendulum is fully constrained; no movement is allowed. We will suppress the third mate *Angle1* before entering *Motion*.

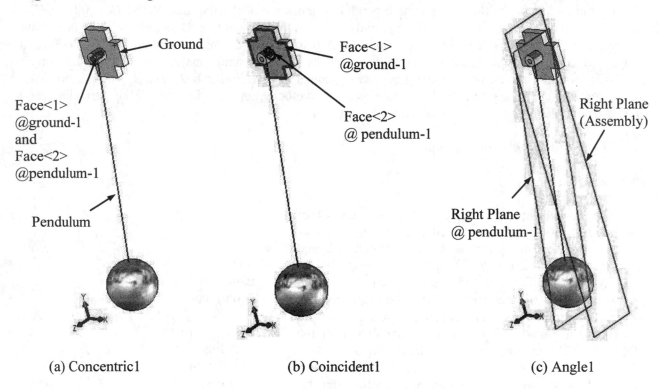

(a) Concentric1                    (b) Coincident1                    (c) Angle1

Figurc 4-2  Assembly Mates Defined in *Lesson4.SLDASM*

### Motion Model

In this motion model, the only moveable body is the pendulum. After suppressing the mate *Angle1*, the pendulum is allowed to rotate about the *Z*-axis of the global coordinate system due to gravity and a small initial angle. Note that the gravity is acting in the negative *Y*-direction. Friction between the pendulum and the ground is assumed zero. You may want to open the assembly file *Lesson4withresults.SLDASM* to play the animation before starting the lesson.

### 4.3  Using *SOLIDWORKS Motion*

Start *SOLIDWORKS* and open assembly file *Lesson4.SLDASM*. First, suppress the third assembly mate *Angle1*.

From the *SOLIDWORKS* browser, expand the *Mates* node, click *Angle1*, and choose *Suppress*. The mate *Angle1* will become inactive. Save the model.

Note that most steps in this lesson are similar to those of *Lesson 3*. You may see some repetitions. Hopefully, with some steps repeated here, you will not have to go back to *Lesson 3* for reference.

Click the *Motion Study* tab (*Motion Study 1*) at the bottom of the graphics area to bring up the *MotionManager* window. Similar to *Lesson 3*, we will add gravity and create a 1.5-second simulation.

### Defining Gravity

Click the *Gravity* button from the *Motion* toolbar to bring up the *Gravity* dialog box. choose *Y* and keep the g-value (9806.65mm/s^2) in the dialog box, as shown in Figure 4-3. In the graphics area, an arrow appears at the right lower corner, pointing downward indicating the direction of the gravity. Click the checkmark on top of the dialog box to accept the gravity.

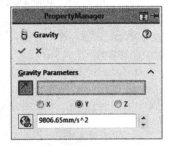

Figure 4-3

### Running Simulation

Choose *Motion Analysis* for the study (see Figure 4-4). Click the *Motion Study Properties* button from the *Motion* toolbar. In the *Motion Study Properties* dialog box (Figure 4-5), enter *300* for *Frames per second*, and click the checkmark. Drag the end time key to 1.5 second mark (see Figure 4-6) in the timeline area to define the simulation duration. Click the *Zoom In* button to zoom in the timeline area.

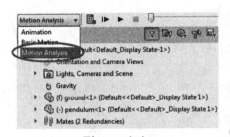

Figure 4-4

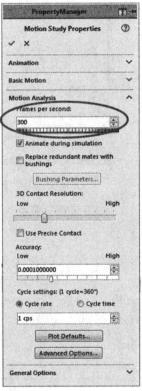

Figure 4-5

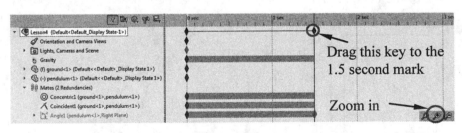

Figure 4-6

Drag this key to the 1.5 second mark

Zoom in

Click *Calculate* button from the *Motion* toolbar to simulate the motion. A 1.5 second simulation will be carried out. You should see that the pendulum starts moving back and forth about the axis of the revolute joint (if not, check if mate *Angle1* is suppressed in *MotionManager* tree). We will graph the position, velocity, and acceleration of the pendulum next.

*Displaying Simulation Results*

The results of angular position, velocity, and acceleration of the pendulum can be directly obtained by right clicking the moving part, *pendulum-1*, from the *MotionManager* tree.

Right-click *pendulum-1*, and choose *Create Motion Plot*. In the *Results* dialog box, choose *Displacement/Velocity/Acceleration*, select *Angular Displacement*, and then *Magnitude*. Click the checkmark.

A graph like that of Figure 4-7 should appear. From the graph, the pendulum swings about the Z-axis between –10 and 10 degrees, as expected (since no friction is involved). Also, it takes about 0.6 seconds to complete a cycle. Note that you can export the graph data, for example, by right clicking the graph and choosing *Export CSV*. Open the spreadsheet and examine the data. From the spreadsheet, the time for the pendulum to swing back to its original position; i.e., –10 degrees, is 0.64 seconds, as shown in the spreadsheet of Figure 4-8. We will carry out calculations to verify these results later. Before we do that, we will graph the angular velocity and acceleration of the pendulum.

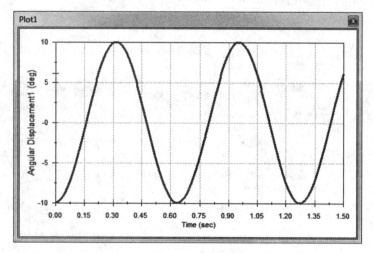

Figure 4-7  Rotation Angle of the Pendulum

Figure 4-8

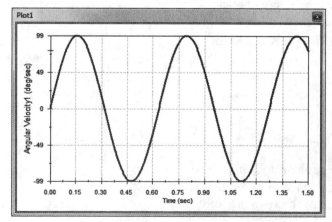 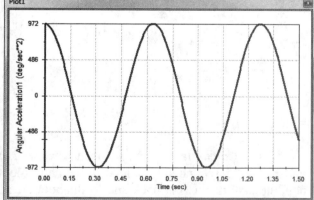

Figure 4-9  Z-Angular Velocity of the Pendulum     Figure 4-10  Z-Angular Acceleration of the Pendulum

Repeat the same steps and in the *Results* dialog box, choose *Angular Velocity* and *Z Component*. Repeat the same steps and choose *Angular Acceleration* and *Z Component*. You should see graphs like those of Figures 4-9 and 4-10. Figure 4-9 shows that the angular velocity starts at *0*, which is expected.

The angular velocity varies roughly between *–100* and *100* degrees/sec. Also, the angular acceleration varies roughly between *–1,000* and *1,000* degrees/sec$^2$. Are these results correct? We will carry out calculations to verify if these graphs are accurate.

Save the model.

## 4.4 Result Verifications

In this section, we will verify the analysis results obtained from *Motion* using particle dynamics theory.

There are four assumptions that we have to make in order to apply the particle dynamics theory to this simple pendulum problem:

(i)   Mass of the rod is negligible (this is why the diameter of the pendulum rod is so small),
(ii)  The sphere is of a concentrated mass,
(iii) Rotation angle is small (remember the initial conditions we defined?), and
(iv)  No friction is present.

The pendulum model has been created to comply with these assumptions as much as possible. We expect that the particle dynamics theory will give us results close to those obtained from simulation. Two approaches will be presented to formulate the equations of motion for the pendulum: energy conservation and Newton's law.

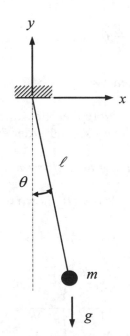

Figure 4-11  Particle Dynamics of Pendulum

### *Energy Conservation*

Referring to Figure 4-11, the kinetic energy and potential energy of the pendulum can be written, respectively, as

$$T = \frac{1}{2}J\dot{\theta}^2 \qquad (4.1)$$

where $J$ is the moment of inertia, i.e., $J = m\ell^2$ in this case; and

$$U = mg\ell\,(1 - \cos\,\theta) \qquad (4.2)$$

According to the energy conservation theory, the total mechanical energy, which is the sum of the kinetic energy and potential energy, is a constant with respect to time; i.e.,

$$\frac{d}{dt}(T + U) = 0 \qquad (4.3)$$

where $t$ represents time. Hence

$$\frac{d}{dt}\left(\frac{1}{2}m\ell^2\dot{\theta}^2 + mg\ell(1 - \cos\theta)\right) = m\ell^2\ddot{\theta} + mg\ell\,\sin\theta = 0 \qquad (4.4)$$

Therefore,

$\ddot{\theta}+\dfrac{g}{\ell}\sin\theta=0$, and

$$\ddot{\theta}+\dfrac{g}{\ell}\theta=0 \tag{4.5}$$

when $\theta\approx0$.

### Newton's Law

From the free-body diagram shown in Figure 4-12, the equilibrium equation of moment at the origin about the $Z$-axis (normal to the paper) can be written as:

$$\sum M = -mg\ell\sin\theta = J\ddot{\theta} = m\ell^2\ddot{\theta} \tag{4.6}$$

Hence

$\ddot{\theta}+\dfrac{g}{\ell}\sin\theta=0$, and

$$\ddot{\theta}+\dfrac{g}{\ell}\theta=0 \tag{4.7}$$

when $\theta\approx0$.

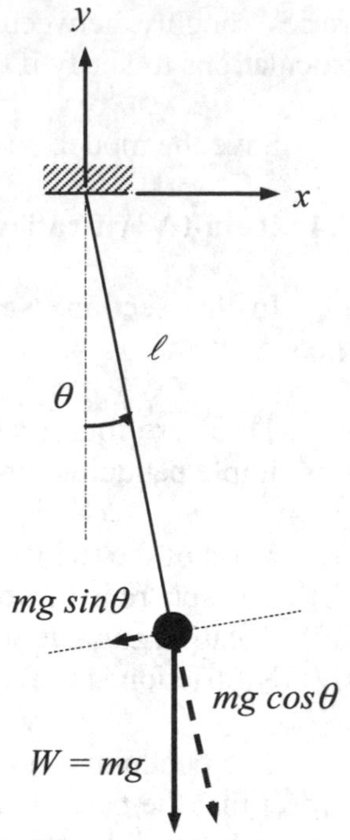

**Figure 4-12  Free Body Diagram**

Note that the same equation of motion has been derived from two different approaches. The linear ordinary second-order differential equation can be solved analytically.

### Solving the Differential Equation

It is well known that the solution of the differential equation is

$$\theta = A_1\cos\omega_n t + A_2\sin\omega_n t \tag{4.8}$$

where $\omega_n = \sqrt{\dfrac{g}{\ell}}$, and $A_1$ and $A_2$ are constants to be determined by initial conditions. Note that

$\omega_n = \sqrt{\dfrac{g}{\ell}} = \sqrt{\dfrac{9806}{100}} = 9.903$ rad/sec, and the natural frequency of the system is $f_n = \omega_n/2\pi = 1.576$ Hz. The period for a complete cycle is $T = 1/f_n = 0.634$ seconds, which is very close to that shown in all graphs (for example, Figure 4-7) as well as the spreadsheet data shown in Figure 4-8 generated using *Export CSV*.

The initial conditions for the pendulum are $\theta(0) = \theta_0 = -10$ degrees, and $\dot{\theta}(0)=0$ degree/sec. Plugging the initial conditions into the solution, we have

$A_1 = \theta_0 = -10$ degrees, and

$A_2 = 0$.

Hence, the solutions are

$$\theta = \theta_0 \cos \omega_n t \qquad (4.9a)$$

$$\dot{\theta} = -\theta_0 \omega_n \sin \omega_n t \qquad (4.9b)$$

$$\ddot{\theta} = -\theta_0 \omega_n^2 \cos \omega_n t \qquad (4.9c)$$

The above equations represent angular position, velocity, and acceleration of the revolute joint, respectively. These equations can be implemented into, for example, *Excel* spreadsheet shown in Figure 4-13, for numerical solutions. Columns B, C, and D in the spreadsheet show the results of Eqs. 4.9a, b, and c, respectively, between *0* and *1.5* seconds with increments of *0.005* seconds. Data in these three columns are graphed in Figures 4-14, 4-15, and 4-16, respectively. Comparing Figures 4-14 to 4-16 with Figures 4-7, 4-9, and 4-10, the results obtained from theory and simulation are very close. The motion model has been properly defined, and *Motion* gives us good results. Note that in the calculation, the angular position of the pendulum is set to zero when it aligns with the vertical axis; therefore, the pendulum swings between *−10* and *10* degrees.

However, even though graphs obtained from *Motion* and spreadsheet calculations are alike these results are not identical. This is because the *Motion* model is not a perfect simple pendulum since mass of the rod is non-zero. If you reduce the diameter of the rod, the *Motion* results should approach those obtained through spreadsheet calculations.

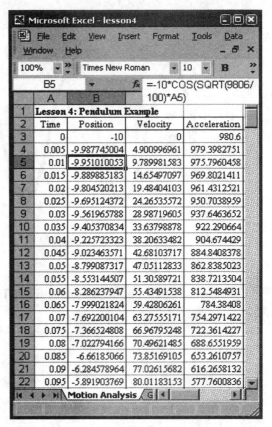

Figure 4-13  The *Excel* Spreadsheet

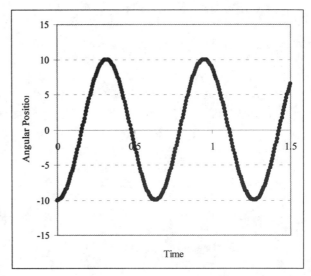

Figure 4-14  Angular Position from Theory

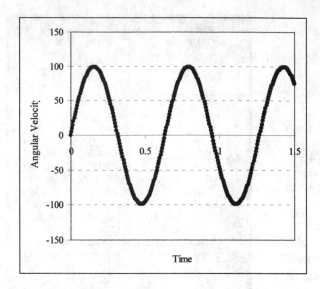

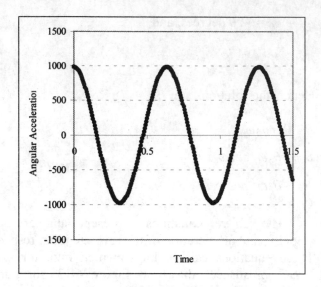

Figure 4-15  Angular Velocity from Theory          Figure 4-16 Angular Acceleration from Theory

**Exercises:**

1.  Create a spring-damper-mass system, as shown in Figure E4-1, using *Motion*. Note that the unstretched spring length is *3* in. The radius of the ball is *0.5* in. and the material is *Cast Alloy Steel* (mass density: *0.2637* $lb_m/in^3$).

    (i)  Find the spring length in the equilibrium condition using *Motion*.

    (ii) Solve the same problem using Newton's laws. Compare your results with those obtained from *Motion*.

2.  If a force $p = 2$ $lb_f$ is applied to the ball as shown in Figure E4-1, repeat both (i) and (ii) of Problem 1.

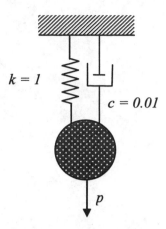

Figure E4-1  The Spring-
Mass-Damper System

**Notes:**

# Lesson 5: A Spring-Mass System

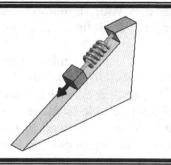

## 5.1  Overview of the Lesson

In this lesson, we will create a simple spring-mass system and simulate its dynamic responses under various scenarios. A schematics of the system is shown in Figure 5-1, in which a steel block of *1"×1"×1"* is sliding along a *30°* slope with a spring connecting it to the top end of the slope. The block will slide back and forth on the slope face under three different scenarios. First, the block will slide due to a small initial displacement, essentially, a free vibration problem. For the second scenario, we will add a friction between the block and the slope face. Finally, we will remove the friction and add a sinusoidal force *p(t)*; therefore, it is a forced vibration problem. Gravity will be turned on for all three scenarios. In this lesson, you will learn how to create the spring-mass model, run a motion analysis, and visualize the analysis results. In addition, you will learn how to add a friction to an assembly mate. The analysis results of the spring-mass example can be verified using particle dynamics theory. Similar to *Lessons 3* and *4* we will formulate the equation of motion, solve the differential equations, graph positions of the block, and compare our calculations with results obtained from *Motion*. Specifically, we will focus on the first and the last scenarios; i.e., free and forced vibrations, respectively.

## 5.2  The Spring-Mass System

### *Physical Model*

Note that the *IPS* units system will be used for this example. The spring constant and the unstretched length (or free length) are $k = 20$ lb$_f$/in. and $U = 3$ in., respectively. As mentioned earlier, the first scenario assumes a free vibration, where the block is stretched *1* in. downward along the *30°* slope. Friction will be imposed for *Scenario 2*, in which the friction coefficient is assumed $\mu = 0.25$. In the third scenario, an external force $p(t) = 10 \cos 360t$ lb$_f$ is applied to the block, as shown in Figure 5-1. All three scenarios assume a gravity of $g = 386$ in/sec$^2$ in the negative *Y*-direction. All three scenarios will be simulated using *Motion*.

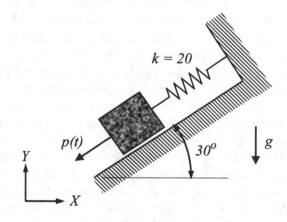

Figure 5-1  The Spring-Mass System

### *SOLIDWORKS Parts and Assembly*

For this lesson, the parts and assembly have been created in *SOLIDWORKS*. There are four model files created: *block.SLDPRT*, *ground.SLDPRT*, *Lesson5.SLDASM*, and *Lesson5withresults.SLDASM*. We will start with *Lesson5.SLDASM*, in which the block is assembled to the ground and no motion entities have been added. The assembly file *Lesson5withresults.SLDASM* contains the complete simulation models with simulation results for the three respective scenarios. You may open this file and review motion of all three scenarios before starting this lesson.

In the assembly models, there are three assembly mates, *Coincident1(ground<1>,block<1>)*, *Coincident3(ground<1>,ball<1>)*, and *LimitDistance1(ground<1>,ball<1>)*, as shown in Figure 5-2. The block is allowed to move along the slope face. You may drag the block in the graphics area; you should be able to move the block on the slope face, but not beyond the slope face. This is because the third mate is defined to restrict the block to move between the lower and upper limits. Choose *Edit > Undo Move Component* from the pull-down menu to restore the block to its neutral position. We will take a look at the assembly mate *LimitDistance1*.

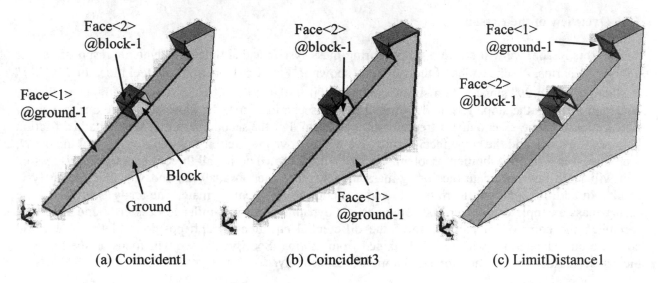

(a) Coincident1                (b) Coincident3                (c) LimitDistance1

Figure 5-2  Assembly Mates Defined in *Lesson5.SLDASM*

From *SOLIDWORKS* browser, click the third mate, *LimitDistance1*, and choose *Edit Feature*. The mate is brought back for reviewing or editing, as shown in Figure 5-3. Note that the distance between the two faces (*Face<2>@block-1* and *Face<1>@ground-1*, see Figure 5-2c) is *3.00* in., which is the neutral position of the block when the spring is undeformed (enter *3* and click the checkmark if the distance is not 3). The upper and lower limits of the distance are *9.00* and *0.00* in., respectively. The length of the slope face is *10* in.; therefore, the upper limit is set to *9.00* in., so that the block will stop when its front lower edge reaches the end of the slope face (since the block width is 1 in.). Note that the *LimitDistance1* is an advanced mate in *SOLIDWORKS*.

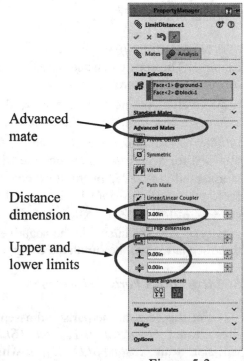

Figure 5-3

### Motion Model

A spring with a spring constant $k = 20$ lb$_f$/in. and an unstretched length $U = 3$ in. will be added to connect the block (*Face<2>*, as shown in Figure 5-2c) with the ground (*Face<1>*). By default, the ends of the spring will connect to the center of the corresponding square faces (see Figure 5-4). No reference points are needed.

This model is adequate to support a free vibration simulation under the first scenario. Note that we will move the block *1* in. downward along the slope face for the simulation. This can be accomplished by changing the distance from *3* to *4* in. in the *LimitDistance1* assembly mate.

Note that there is no need to suppress any mate before entering *Motion*. As mentioned earlier, a friction force will be added between the block and the slope face for *Scenario 2*. In addition, a sinusoidal force $p(t) = 10 \cos 360t$ will be added to the block for the 3rd scenario. Gravity will be turned on for all three scenarios.

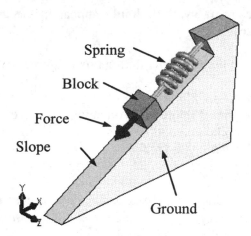

Figure 5-4 Spring Mass Dynamic Model

### 5.3 Using *SOLIDWORKS Motion*

Start *SOLIDWORKS* and open assembly file *Lesson5.SLDASM*.

Before creating any motion entities, always check the units system. Similar to *Lesson 3*, choose from the pull-down menu

*Tools > Options*

and choose the *Document Properties* tab in the *Document Properties-Units* dialog box, then click the *Units* node. The *IPS* should have been selected. Close the dialog box. In this units system, the gravity is *386* in/sec$^2$ in the negative *Y*-direction of the global coordinate system.

Click the *Motion Study* tab (*Motion Study 1*) at the bottom of the graphics area to bring up the *MotionManager* window.

### Defining Gravity

Click the *Gravity* button 🔅 from the *Motion* toolbar to bring up the *Gravity* dialog box. Choose *Y* and keep the g-value (386.09in/s^2) in the dialog box, as shown in Figure 5-5. In the graphics area, an arrow appears at the right lower corner ⬇, pointing downward indicating the direction of the gravity.

Figure 5-5

Click the checkmark ✅ on top of the dialog box to accept the gravity. A gravity node (*Gravity*) should appear in the *MotionManager* tree.

### Defining Spring

Click the *Spring* button 📊 from the *Motion* toolbar to bring up the *Spring* dialog box. In the *Spring* dialog box (Figure 5-6), choose *Linear Spring* (default). The empty field right underneath the *Spring parameters* label is active for you to pick entities to define ends of the spring. Pick the face at top right of the ground (see Figure 5-7). Rotate the view, and then pick the face in the block, as shown in Figure 5-7.

A spring symbol should appear in the graphics area, connecting the center points of the two selected faces.

In the *Spring* dialog box (Figure 5-6), enter the followings:

*Stiffness: 20*
*Length: 3* (Note that you have to deselect the *Update to model changes* box before entering this value)
*Coil Diameter: 0.75*
*Number of Coils: 5*
*Wire Diameter: 0.25*

Click the checkmark to accept the definition and close the *Spring* dialog box. A spring node (*LinearSpring1*) should appear in the *MotionManager* tree.

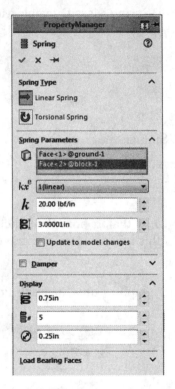

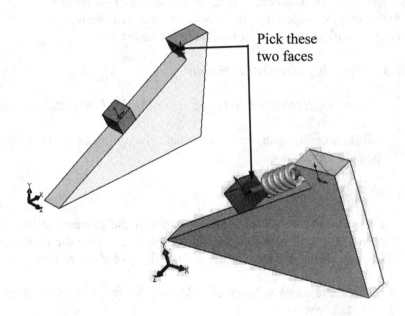

Figure 5-7  Picking Two Faces to Define the Spring

Figure 5-6  The *Spring* Dialog Box

### *Defining Initial Position*

We would like to stretch the spring *1* in. downward along the slope face as the initial position for the block. The block will be released from this position to simulate a free vibration; i.e., the first scenario. We will edit the mate *LimitDistance1*; enter *4* for the distance dimension.

From the *SOLIDWORKS* browser, expand the *Mates* node listed in the browser, click *LimitDistance1(ground<1>,ball<1>)*, and choose *Edit Feature* . You should see the definition of the assembly mate in the dialog box like that of Figure 5-3. Enter *4.00* in. for the distance, and click the checkmark on top to accept the change. In the graphics area, the block should move to the *4.00* in. position on the slope face.

### Defining and Running Simulation

Choose *Motion Analysis* for the study. Click the *Motion Study Properties* button ⚙ from the *Motion* toolbar. In the *Motion Study Properties* dialog box, enter *500* for *Frame per second*, and click the checkmark on top. Zoom in the timeline area until you can see tenth second marks. Drag the end time key to 0.25 second mark (see Figure5-8) in the timeline area to define the simulation duration.

Click *Calculate* button 🗔 from the *Motion* toolbar to simulate the motion. A 0.25 second simulation will be carried out.

If you see a warning message stating *The playback speed or frames per second settings for the motion study will result in poor performance given the current motion study duration.* Just ignore it (select *No*) since 0.25 seconds will be sufficient for our analysis.

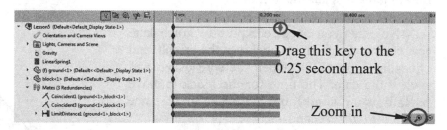

Figure 5-8

After a few seconds, you should see the block start moving back and forth on the slope face. You may want to adjust the playback speed to 10% to slow down the animation (by simply choosing *0.1x* from the scroll-down menu next to the animation slider) since the total simulation duration is only 0.25 seconds.

Next, we will graph the spring length which describes the displacement of the block on the slope face (instead of *X*- or *Y*-component). This displacement graph should reveal a sinusoidal function as we have seen in many vibration examples of *Physics*.

### Displaying Simulation Results

Click the *Results and Plots* button 🖳 from the *Motion* toolbar. In the *Results* dialog box (refer to Figure 5-9), choose *Displacement/Velocity/Acceleration*, select *Linear Displacement*, and then *Magnitude*. Similar to the spring, pick the face at top right of the ground (see Figure 5-7). Rotate the view, and then pick the face in the block, as shown in Figure 5-7. Click the checkmark to accept the graph.

Figure 5-9  The *Results* Dialog Box

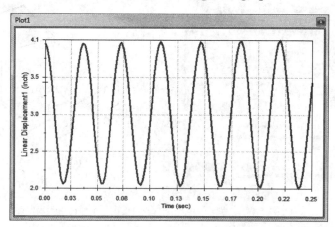

Figure 5-10  The Displacement Graph: *Scenario 1*

A graph like that of Figure 5-10 should appear. From the graph, the *block* moves on top of the slope face between *2* and *4* in. This is because the unstretched length of the spring is *3* in. and we stretched the spring *1* in. to start the motion. Also, it takes about *0.04* seconds to complete a cycle, which is small. The small vibration period can be attributed to the fact that the spring is fairly stiff (*20* lb$_f$/in).

Note that if you see a graph similar to Figure 5-10 but with upper and lower bounds 4.5 and 2.5 in., respectively, you may consider selecting two vertices (Figure 5-11), instead of faces, for defining the displacement graph.

Note that you can also export the graph data, for example, by right clicking the graph and choosing *Export CSV*. Open the spreadsheet and review the data. The time for the block to move back to its initial position; i.e., when the distance is *4* in., is *0.037* seconds.

We will carry out calculations to verify these results later in Section 5.4. Before we do that, we will work on two more scenarios: with friction and with the addition of an external force.

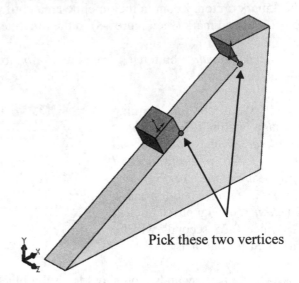

Pick these two vertices

Figure 5-11  Pick Vertices for Plot

Double click the *Motion Study 1* tab, and change the name to *Scenario 1*. Save the model before moving to the next scenario.

### Scenario 2: With Friction

We will add a friction force to the mate *Coincident1* between the block and the ground. The friction coefficient is $\mu = 0.25$.

Duplicate the motion study by right clicking the *Scenario 1* tab and choosing *Copy Study* (see Figure 5-12). A new motion study with a default name *Motion Study 1* will appear. Change its name to *Scenario 2*.

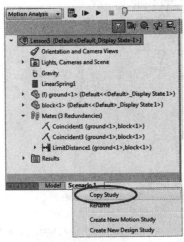

Figure 5-12

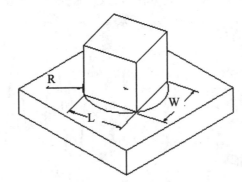

Figure 5-13

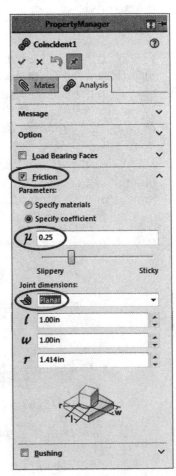

Figure 5-14

Note that for calculating friction effects, *Motion* models the coincident mate between two faces as a planar joint, where one block slides and rotates on the surface of another block, as illustrated in Figure 5-13, where *L* is the length of the top (sliding) rectangular block, *W* is the width of the top rectangular block, and *R* is the radius of a circle, centered at the center of the top block face in contact with the bottom block, which circumscribes the face of the sliding block.

For this example, we will set *L* and *W* to *1*, and *R* to *1.414*; i.e., $\sqrt{2}$. We will add friction from *SOLIDWORKS* browser.

The friction coefficient can be entered through the 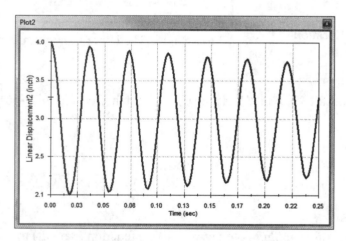 *Analysis* tab of the mate dialog box. From the *SOLIDWORKS* browser, expand the *Mates* node, click the *Coincident1* node and choose *Edit Feature* . In the *Coincident1* dialog box (Figure 5-14), choose the *Analysis* tab, choose *Friction*, enter *0.25* for *Coefficient (μ)*, choose *Planar* for joint, and enter *Length: 1*, *Width: 1*, and *Radius: 1.414*. Click checkmark on top of the dialog box to accept the definition. Click it one more time to close the box.

Run a simulation (with the same simulation parameters as those of *Scenario 1*). Graph the displacement of the block; you should see a graph similar to that of Figure 5-15. The amplitude of the graph (that is, the distance the block travels) is decreasing over time due to friction. Save the model. We will move into *Scenario 3*.

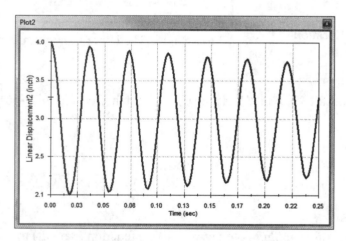

Figure 5-15  The Displacement Graph: *Scenario 2*

### Scenario 3: With External Force p(t) but No Friction

In this scenario we will add an external force $p(t) = 10 \cos 360t$ at the center of the end face of the *block* in the downward direction along the slope face. At the same time, we will remove the friction in order to simplify the problem.

Duplicate the motion study and change its name to *Scenario 3*.

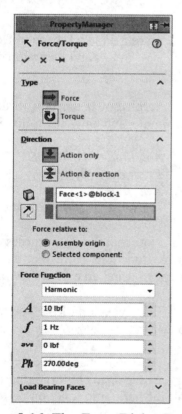

Figure 5-16  The *Force* Dialog Box

From the *SOLIDWORKS* browser, right click the *Coincident1* node to edit its definition. In the *Coincident1* dialog box, choose *Analysis* tab, and deselect the *Friction* by clicking the small box in front of it. All parameters and selections on the dialog box should become inactive. Click checkmark to accept the definition.

The force can be added by choosing the *Force* button ↖ from the *Motion* toolbar. In the *Force* dialog box (Figure 5-16), choosing *Linear Force* for *Force Type* and *Action Only* for *Direction*, and then pick the end face of the block, as shown in Figure 5-17, for the location of the applied force.

The picked face *Face<1>@block-1* is now listed. An arrow appears in the downward direction (shown in Figure 5-17), indicating the direction of the force. That is, the force will be applied to the center of the end face and in the downward direction that is normal to the selected face.

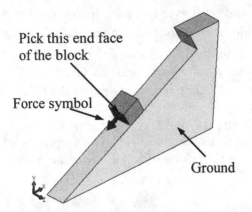

Figure 5-17  Pick Face to Defining Location and Direction of the Force

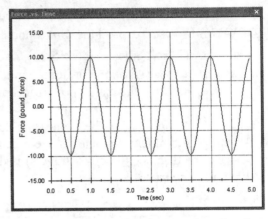

Figure 5-18  The Force Function Graph

Scroll down the box to define *Force Function*. Click the *Function* tab (see Figure 5-16), choose *Harmonic*, and enter the following:

*Amplitude (A): 10 lbf*
*Frequency (f): 1 Hz*
*Ave: 0 lbf*
*Phase Shift (Ph): 270 degrees*

Note that the *270* degrees entered for *Phase Shift* is to convert a sine function (default) to the desired cosine function. Note that the force defined is a sinusoidal function like that shown in Figure 5-18 (graphed using an earlier *Motion* version).

Click checkmark ✓ on top of the dialog box to accept the definition. A force node (*Force1*) should appear in the *MotionManager* tree.

Next, we change the simulation duration to *0.5* seconds (in order to see a graph later that covers a larger time span) by dragging the end time key to 0.5-second mark. Note that the *0.5*-second duration is half the harmonic function period of the force applied to the block.

Click the *Calculate* button 🖩 to run a simulation (with the same simulation parameters as those of *Scenario 1*, except with duration extended to 0.5 seconds). After a short moment, the block starts moving.

As soon as the simulation is completed, a graph like that of Figure 5-19 should appear. From the graph, the *block* moves on the slope face roughly for *2* in. back and forth (since friction is turned off). The vibration amplitude is enveloped by a cosine function due to the external force *p(t)*. Also, it takes about

*0.04* seconds to complete a cycle, which is unchanged from the previous cases. We will carry out calculations to verify these results later. Save the model. Note that if you see a graph that shows a sinusoidal curve similar to that of Figure 5-19, but with increasing amplitude, the result is incorrect. One possible remedy is to increase the accuracy to 0.00001 (default is 0.0001) from the *Motion Study Properties* dialog box, as shown in Figure 5-20, by clicking the *Motion Study Properties* button ⚙ from the *Motion* toolbar.

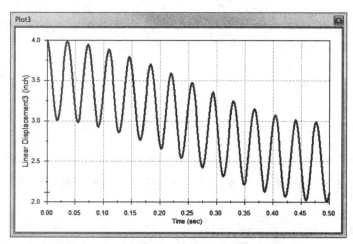

Figure 5-19  The Displacement Graph: *Scenario 3*

### 5.4   Result Verifications

In this section, we will verify analysis results of *Scenarios 1* and *3* obtained from *Motion*. We will assume that the *block* is of a concentrated mass so that the particle dynamics theory is applicable.

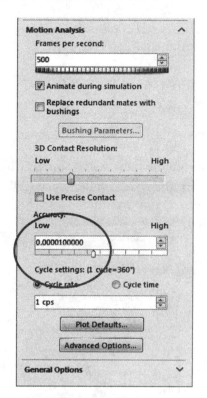

Figure 5-20

We will start with *Scenario 1* (i.e., free vibration with gravity), and then solve the equations of motion for *Scenario 3* (forced vibration, no friction).

***Equation of Motion: Scenario 1***

From the free-body diagram shown in Figure 5-21, applying Newton's Second Law and force equilibrium along the *X*-direction (i.e., along the 30° slope), we have

$$\sum F_x = mg\sin\theta - k(x-U) = m\ddot{x} \qquad (5.1)$$

Therefore,

$$m\ddot{x} + k(x-U) = mg\sin\theta \qquad (5.2)$$

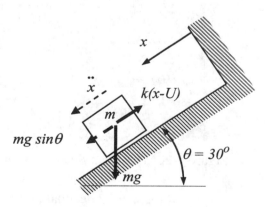

Figure 5-21  The Free-Body Diagram

where *m* is the mass of the block, *U* is the unstretched length of the spring, *x* is the distance between the mass center of the block and the top right end of the slope face, measured from the top right end.

The double dots on top of $x$ represent the second derivative of $x$ with respect to time. Rearranging Eq. 5.2, we have

$$m\ddot{x} + kx = mg \sin\theta + Uk \qquad (5.3)$$

where both terms on the right are time-independent.

This is a second-order ordinary differential equation. It is well known that the general solution of the differential equation is

$$x_g = A_1 \cos\omega_n t + A_2 \sin\omega_n t \qquad (5.4)$$

where $\omega_n = \sqrt{\dfrac{k}{m}}$, and $A_1$ and $A_2$ are constants to be determined with initial conditions. Note that the mass of the steel block is *0.264 pounds* (this is pound mass, $lb_m$). This can be obtained from *SOLIDWORKS* by selecting the block, and choosing, from the pull-down menu, *Tools > Evaluate > Mass Properties* from the *Mass Properties* dialog box (Figure 5-22).

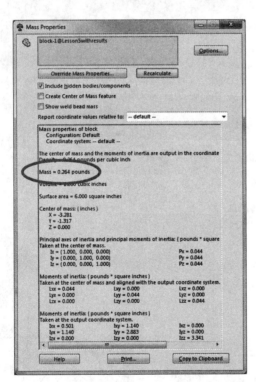

Figure 5-22  The *Mass Properties* Dialog Box

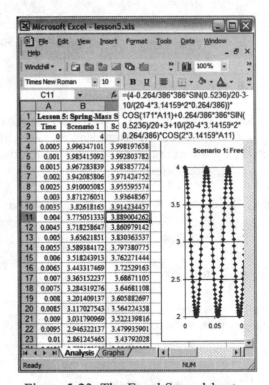

Figure 5-23  The Excel Spreadsheet

Note that there are *2* decimal points set in *SOLIDWORKS* by default. You may increase it using the *Document Properties – Units* dialog box (choose from pull-down menu *Tools > Options*).

Note that the pound-mass unit $lb_m$ is not as common as the slug that we are more familiar with. The corresponding force unit of $lb_m$ is $lb_m$ in/sec$^2$ according to Newton's Second Law.

Also, a $1$-$lb_m$ mass block weighs $1$ $lb_f$ on earth; therefore, $1$ $lb_f = 386$ $lb_m$ in/sec². The stiffness we employed for the spring becomes $k = 20$ $lb_f$/in $= 20 \times 386$ $lb_m$ in/sec². Hence,

$$\omega_n = \sqrt{\frac{k}{m}} = \sqrt{\frac{20 \times 386}{0.264}} = 171.0 \text{ rad/sec} = 9{,}798 \text{ degrees/sec, and the natural frequency of the system is } f_n$$

$= \omega_n/2\pi = 27.2$ Hz. The period for a complete vibration cycle is $T = 1/f_n = 0.0367$ seconds, which is very close to what was obtained by reviewing the graph (for example, the graph shown in Figure 5-10) and spreadsheet data converted from the graph. For more details regarding the force and mass in the English units system, please refer to Appendix B. Appendix B should clarify some of the confusion you might have in $lb_m$ and $lb_f$.

The particular solution of Eq. 5.3 is

$$x_p = \frac{mg \sin\theta}{k} + U \tag{5.5}$$

Therefore, the total solution is

$$x = x_g + x_p = A_1 \cos\omega_n t + A_2 \sin\omega_n t + \frac{mg \sin\theta}{k} + U \tag{5.6}$$

The initial conditions for the spring-mass system are $x(0) = x_0 = 4$ in., and $\dot{x}(0) = 0$ in/sec. Plugging the initial conditions into Eq. 5.6 (you will have to take the derivative of Eq. 5.6 for $\dot{x}(0)$), we have

$$A_1 = x_0 - \left(\frac{mg \sin\theta}{k} + U\right), \text{ and}$$

$$A_2 = 0.$$

Hence, the overall solution is

$$x = \left(x_0 - \frac{mg \sin\theta}{k} - U\right)\cos\omega_n t + \frac{mg \sin\theta}{k} + U \tag{5.7}$$

Equation 5.7 can be implemented into Microsoft® Excel spreadsheet, as shown in Figure 5-23. Column B in the spreadsheet shows the results of Eq. 5.7, which is graphed in Figure 5-24. Comparing Figure 5-24 with Figure 5-10, the results obtained from theory and *Motion* agree very well, which means the motion model has been properly defined, and *Motion* gives us good results.

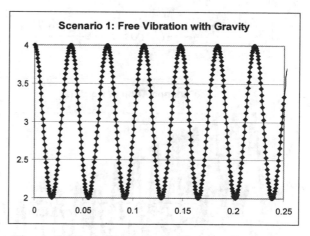

Figure 5-24 Solution from Theory: *Scenario 1*

### Equation of Motion: Scenario 3

Refer to the free-body diagram shown in Figure 5-21 again. For *Scenario 3* we must include the force $p = f_0 \cos(\omega t)$ along the $X$-direction for force equilibrium; i.e.,

$$m\ddot{x} + k(x - U) = mg \sin\theta + f_0 \cos(\omega t) \tag{5.8}$$

where $f_0 = 10$ lb$_f$ and $\omega = 360$ degree/sec$= 2\pi$ rad/sec. Rearranging Eq. 5.8, we have

$$m\ddot{x} + kx = mg\,\sin\theta + Uk + f_0\,\cos(\omega t) \tag{5.9}$$

where the right-hand side consists of constant and time-dependent terms.

For the constant terms, the particular solution is identical to that of *Scenario 1*; i.e., Eq. 5.5. For the time-dependent term, $p = f_0\,\cos(\omega t)$, the particular solution is

$$x_{p_2} = \frac{f_0}{k - \omega^2 m}\cos\omega t \tag{5.10}$$

Therefore, the overall solution of Eq. 5.9 becomes

$$x = x_g + x_p + x_{p_2} = A_1\cos\omega_n t + A_2\sin\omega_n t + \frac{mg\,\sin\theta}{k} + U + \frac{f_0}{k - \omega^2 m}\cos\omega t \tag{5.11}$$

Plugging the initial conditions into the solution, we have

$$A_1 = x_0 - \left(\frac{mg\,\sin\theta}{k} + U + \frac{f_0}{k - \omega^2 m}\right), \text{ and } A_2 = 0.$$

Hence, the complete solution is

$$x = \left(x_0 - \frac{mg\,\sin\theta}{k} - U - \frac{f_0}{k - \omega^2 m}\right)\cos\omega_n t + \frac{mg\,\sin\theta}{k} + U + \frac{f_0}{k - \omega^2 m}\cos\omega t$$
$$= \left[\left(x_0 - \frac{mg\,\sin\theta}{k} - U\right)\cos\omega_n t + \frac{mg\,\sin\theta}{k} + U\right] + \frac{f_0}{k - \omega^2 m}\left(\cos\omega t - \cos\omega_n t\right) \tag{5.12}$$

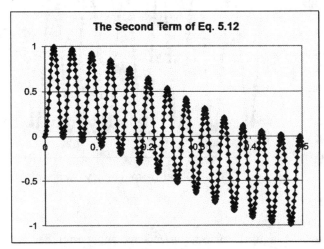

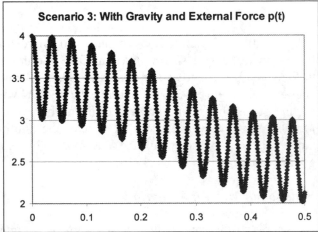

Figure 5-25  Graph of the Second Term of Eq. 5.12

Figure 5-26  Solution from Theory: *Scenario 3*

Note that terms grouped in the first bracket of Eq. 5.12 are identical to those of Eq. 5.7; i.e., *Scenario 1*. The second term of Eq. 5.12 graphed in Figure 5-25 represents the contribution of the external force *p(t)* to the block motion. The graph shows that the amplitude of the block is kept within *1* in., but the position of the block varies in time. The vibration amplitude is enveloped by a cosine function.

The overall solution of *Scenario 3*; i.e., Eq. 5.12, is a combination of graphs shown in Figures 5-24 and 5-25. In fact, Eq. 5.12 has been implemented in Column C of the spreadsheet. The data are graphed in Figure 5-26.

Comparing Figure 5-26 with Figure 5-19, the results obtained from theory and *Motion* are very close. Note that the spreadsheet shown in Figure 5-23 can be found at the publisher's website (filename: *lesson5.xls*).

**Exercises:**

1. Show that Eq. 5.12 is the correct solution of *Scenario 3* governed by Eq. 5.8 by simply plugging Eq. 5.12 into Eq. 5.8.

2. Repeat the *Scenario 3* of this lesson, except changing the external force to $p(t) = 10 \cos 9798.0t$ lb$_f$. Will this external force change the vibration amplitude of the system? Can you simulate this resonance scenario in *Motion*?

3. Add a damper with damping coefficient $C = 0.01$ lb$_f$ sec/in. and repeat the *Scenario 1* simulation using *Motion*.

    (i) Calculate the natural frequency of the system and compare your calculation with that of *Motion*.

    (ii) Derive and solve the equations that describe the position and velocity of the mass. Compare your solutions with those obtained from *Motion*.

# Lesson 6: A Slider-Crank Mechanism

## 6.1 Overview of the Lesson

In this lesson, we will learn how to create simulation models for a slider-crank mechanism and conduct three analyses: kinematic, interference, and dynamic. You will learn how to select assembly mates to connect parts in order to create a successful motion model. We will first drive the mechanism by rotating the crank with a constant angular velocity; basically, conducting a kinematic analysis. After we complete a kinematic analysis, we will turn on interference checking and repeat the analysis to see if parts interfere or collide. It is very important to make sure no interference exists between parts while the mechanism is in motion. Finally, we will go over a dynamic analysis where we will add a firing force to the piston. This lesson will start with a brief overview about the slider-crank assembly created in *SOLIDWORKS*. At the end of this lesson, we will verify the kinematic simulation results using theory and computational methods commonly found in mechanism design textbooks.

## 6.2 The Slider-Crank Example

### *Physical Model*

The slider-crank mechanism is essentially a four-bar linkage, as shown in Figure 6-1. They are often found in mechanical systems; e.g., internal combustion engine and oil-well drilling equipment. For the internal combustion engine, the mechanism is driven by a firing load that pushes the piston, converting the reciprocal motion into rotational motion at the crank.

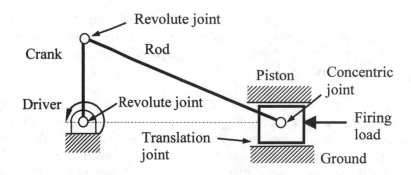

Figure 6-1  Schematic View of the Slider-Crank Mechanism

In the oil-well drilling equipment, a torque is applied at the crank. The rotational motion is converted to a reciprocal motion at the piston that digs into the ground. Note that in any case the length of the crank must be smaller than that of the rod in order to allow the mechanism to operate. This is commonly referred to as Grashof's law. In this example, the lengths of the crank and rod are 3 in. and 8 in., respectively.

Note that the units system chosen for this example is *IPS* (*in-lb_f-sec*). All parts are made of Aluminum, *2014 Alloy*. No friction is assumed between any pair of the components (parts or subassemblies).

### SOLIDWORKS Parts and Assembly

The slider-crank system consists of five parts and one subassembly. They are bearing, crank, rod, pin, piston, and rodandpin (subassembly, consisting of rod and pin). An explode view of the mechanism is shown in Figure 6-2. *SOLIDWORKS* parts and assembly have been created for you. They are *bearing.SLDPRT*, *crank.SLDPRT*, *rod.SLDPRT*, *pin.SLDPRT piston.SLDPRT*, and *rodandpin.SLDASM*. In addition, there are two assembly files: *Lesson6.SLDASM* and *Lesson6withresults.SLDASM*. You can find these files at the publisher's web site.

We will start with *Lesson6.SLDASM*, in which all components are properly assembled. In this assembly the bearing is anchored (ground) and all other parts are fully constrained. We will suppress one assembly mate in order to allow for movement.

Same as before, the assembly file *Lesson6withresults.SLDASM* consists of a complete simulation model with simulation results. You may want to open this file to see the motion simulation of the mechanism. In the assembly file with complete simulation results, a mate (*Concentric2*) has been suppressed. You can also see how the parts move by moving the mouse pointer to the graphics area and dragging a movable part, for example the crank. The whole mechanism will move accordingly.

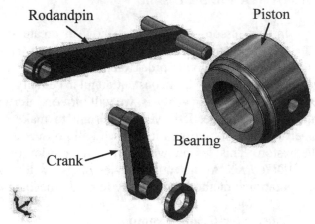

Figure 6-2  Slider-Crank Assembly (Explode View)

There are eight assembly mates, including five coincident and three concentric, defined in the assembly. You may want to expand the *Mates* node in the browser to see the list of mates. Click any of the mates; you should see the geometric entities selected for the assembly mate highlighted in the graphics area.

The first three mates (*Concentric1*, *Coincident1*, and *Coincident2*) assemble the crank to the fixed bearing, as shown in Figure 6-3a. As a result, the crank is completely fixed. Note that the mate *Coincident2* orients the crank to the upright position, defining the initial configuration of the mechanism. This mate will be suppressed before entering *Motion*. Suppressing this mate will allow the crank to rotate on the bearing. Note that combining these two mates, *Concentric1* and *Coincident1*, yields a revolute joint. However, in *Motion 2017* only assembly mates are employed for motion models. Kinematic joints, such as revolute joint, cylindrical joint, etc. commonly discussed in kinematic and dynamic textbooks, cannot be found in *Motion 2017*. If interested, you may refer to Appendix A for more discussion on the subject of kinematic joints.

The next two mates (*Concentric2* and *Coincident3*) assemble the rod to the crank, as shown in Figure 6-3b. Unlike the crank, the rod is allowed to rotate with respect to the crank. The next two mates (*Concentric3* and *Coincident4*) assemble the piston to the pin, allowing the piston to rotate about the pin. The final mate (*Coincident5*) eliminates the rotation by mating two planes, *Right Plane* of the piston and the *Top Plane* of the assembly, as shown in Figure 6-3c.

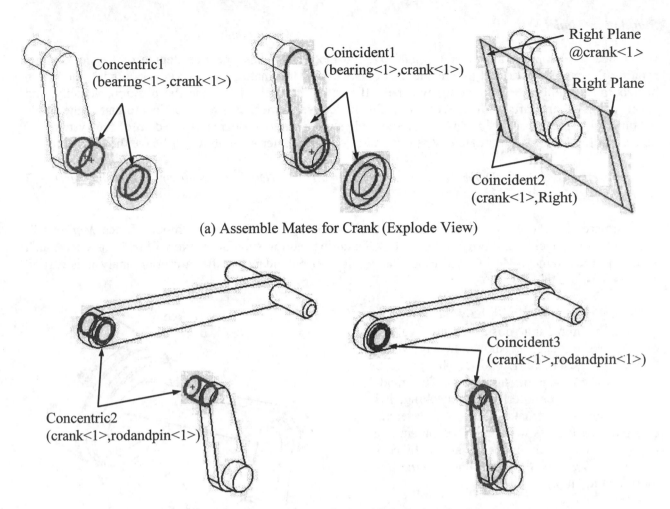

Concentric1
(bearing<1>,crank<1>)

Coincident1
(bearing<1>,crank<1>)

Right Plane
@crank<1>

Right Plane

Coincident2
(crank<1>,Right)

(a) Assemble Mates for Crank (Explode View)

Coincident3
(crank<1>,rodandpin<1>)

Concentric2
(crank<1>,rodandpin<1>)

(b) Assemble Mates for Rod (Explode View)

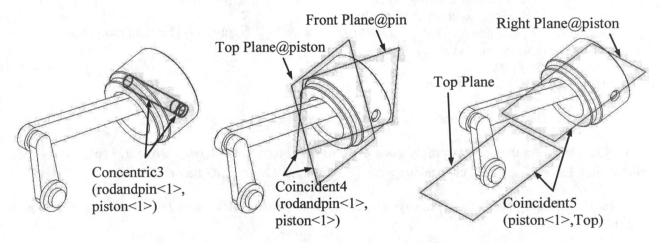

Front Plane@pin

Top Plane@piston

Right Plane@piston

Top Plane

Concentric3
(rodandpin<1>,
piston<1>)

Coincident4
(rodandpin<1>,
piston<1>)

Coincident5
(piston<1>,Top)

(c) Assemble Mates for Piston (Unexplode View)

Figure 6-3  Assembly Mates Defined for the Mechanism

*Simulation Model*

Once we suppress the mate *Coincident2*, the mechanism has one free degree of freedom, either rotating the crank or the rod (about the axis of the concentric mates; i.e., Z-axis), or moving the piston horizontally (along the X-direction) will be sufficient to uniquely determine the position, velocity, and acceleration of any parts in the mechanism. Kinematically this mechanism is identical to that of the single piston engine presented in *Lesson 1*. We know physically the total degree of freedom of the system is *1* before adding a rotary motor. However, if we carry out a Gluebler's count, we will have the following:

*3 (moving bodies) × 6 (DOFs/body) – 3 (concentric) × 4 (DOFs/concentric) – 4 (coincident) × 3 (DOFs/coincident) = 18 –24= –6*

Apparently, there are seven redundant DOFs embedded in the motion model. Since *Motion* will automatically detect and remove redundant DOFs during motion simulation, we will not worry too much about the redundancy. Motion will report the number of redundancies after a motion analysis is carried out. We will see this later.

The mechanism will be first driven by rotating the crank at a constant angular velocity of *360* degrees/sec, as shown in Figure 6-4. Gravity will be turned off. This will be essentially a kinematic simulation. This model will also serve for interference checking. It is very important to make sure no interference exists between parts while the mechanism is in motion. The simulation results are included in *Lesson6withresults.SLDASM* under *Kinematic* motion study tab.

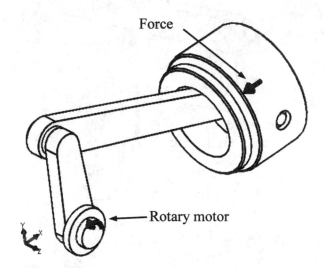

The next and final analysis will be dynamic, where we will add a firing force to the piston for a dynamic simulation. The results are included in *Lesson6withresults.SLDASM* under *Dynamic* motion study tab.

Figure 6-4  The Simulation Model

### 6.3  Using *SOLIDWORKS Motion*

Start *SOLIDWORKS* and open assembly file *Lesson6.SLDASM*.

First, suppress the third assembly mate *Coincident2*. From the *Assembly* browser, expand the *Mates* node, click *Coincident2*, and choose *Suppress* ⬇. The mate *Coincident2* becomes inactive.

Before creating any entities, always check the units system. Make sure *IPS* units system is chosen for this example.

Click the *Motion Study* tab at the bottom of the graphics area to bring up the *MotionManager* window. Double click the tab and rename the study *Kinematic*.

We will add a rotary motor to drive the crank and conduct a 1-second kinematic analysis. Later we will add a force for a dynamic simulation.

### Adding a Rotary Motor

Click the *Motor* button ![motor icon] from the *Motion* toolbar to bring up the *Motor* dialog box (Figure 6-5). Choose *Rotary Motor* (default). Move the pointer to the graphics area, and pick a circular arc to define the rotation direction of the rotary motor; e.g., the circular edge on the lower shaft of the crank, as shown in Figure 6-6. A circular arrow appears indicating the rotational direction of the rotary motor. A counterclockwise direction (Z-direction) is desired. You may change the direction by clicking the direction button ![direction icon] right under *Component/Direction*. Choose *Constant speed* and enter *60* RPM for speed. Click the checkmark ![checkmark] on top of the dialog box to accept the motor definition. You should see a rotary motor *RotaryMotor1* added to the *MotionManager* tree.

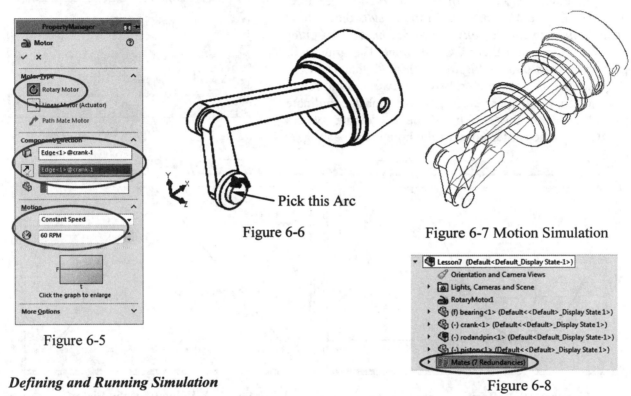

Figure 6-5

Figure 6-6

Figure 6-7 Motion Simulation

Figure 6-8

### Defining and Running Simulation

Choose *Motion Analysis* for the study. Click the *Motion Study Properties* button ![settings icon] from the *Motion* toolbar. In the *Motion Study Properties* dialog box, enter *200* for *Frames per second*, and click the checkmark on top. Drag the end time key to one second mark in the timeline area to define the simulation duration. Zoom in to the timeline area if needed.

Click the *Calculate* button ![calculate icon] from the *Motion* toolbar to simulate the motion. A one second simulation will be carried out. After a short moment, you should see that the mechanism starts moving. The crank rotates *360* degrees as expected and the piston moves forward and backward for a complete cycle, similar to that of Figure 6-7, since the default simulation duration is *1* second.

Note also the *Mates* in the *MotionManager* tree, where number of redundancies is *7* (see Figure 6-8), as expected. Right click the *Mates* node and choose *Degrees of Freedom* to show details of the redundancies (Figure 6-9), in which the redundant degrees of freedom removed by *Motion* are listed. Note that no reaction forces are recorded for the joints at the redundant degree of freedom.

If you encounter a problem in conducting a motion analysis, try to run a *Basic Motion* analysis first. Occasionally, *Motion Analysis* may have problems in addressing redundant constraints; *Basic Motion* analysis seems to clear up the problem.

### Saving and Reviewing Results

We will create four graphs for the mechanism: *X*-position, *X*-velocity, and *X*-acceleration at the mass center of the piston, and angular velocity of mate *Concentric2* (between crank and rod).

Right click *piston<1>* in the *MotionManager* tree, and choose *Create Motion Plot*. In the *Results* dialog box, choose *Displacement/Velocity/Acceleration*, select *Center of Mass Position*, and then *X Component*. The graph will appear similar to that of Figure 6-10. Note that from the graph, the piston mass center moves between about *5* and *11* in. horizontally, in reference to the global coordinate system, in which the origin of the coordinate system coincides with the center point of the hole in the bearing.

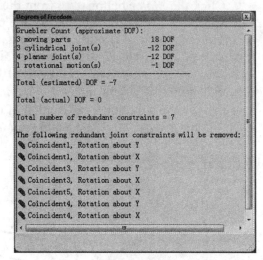

Figure 6-9

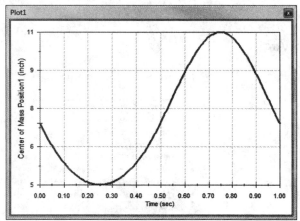

Figure 6-10  *X*-Position of the Piston

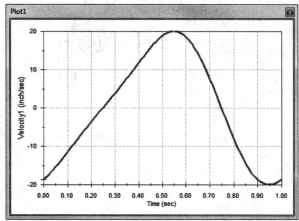

Figure 6-11  *X*-Velocity of the Piston

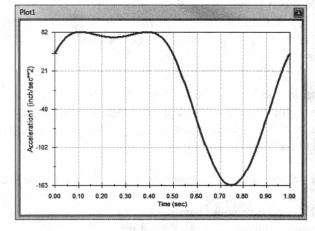

Figure 6-12  *X*-Acceleration of the Piston

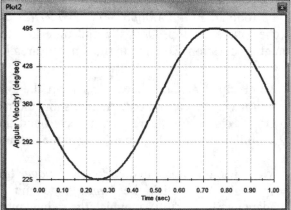

Figure 6-13  Angular Velocity of *Concentric2*

At the starting point, the crank is at the upright position, and the piston is located at *7.42* in. (that is, $\sqrt{8^2 - 3^2}$ ) to the right of the origin of the global coordinate system. Note that the lengths of the crank and rod are *3* and *8* in., respectively. When the crank rotates to *90* degrees counterclockwise, the position becomes *5* (which is *8–3=5*). When the crank rotates *270* degrees, the piston position is *11* (which is *8+3*).

Repeat the same steps to create graphs for the velocity (choose *Linear Velocity*) and acceleration (choose *Linear Acceleration*) of the piston mass center in the *X*-direction. The graphs should be similar to those of Figures 6-11 and 6-12, respectively.

The graph of the angular velocity of the mate *Concentric2 (crank<1>, rodandpin<1>)* can be created by right clicking the mate in the *MotionManager* tree, choosing *Creating Motion Plot*, and selecting *Displacement/Velocity/Acceleration*, *Angular Velocity*, and then *Z Component* in the *Results* dialog box. The graph should be similar to that of Figure 6-13.

If you expand the *Results* node in the *MotionManager* tree, you should see there are four graphs listed, *Plot1<Center of Mass Position1>*, *Plot2<Velocity1>*, *Plot3<Acceleration1>*, and *Plot4<Angular Velocity1>*, as shown in Figure 6-14.

### Interference Check

Next we will learn how to perform interference checking. *Motion* allows you to check for interference in your mechanism as components move. In addition to the interference checking in *Motion*, you can detect interference between components in the *SOLIDWORKS* assembly model, regardless of whether a component participates in the motion model.

Using the interference checking capability in *Motion*, you can find the interference that occurs between the selected components as the mechanism moves through a specified duration of motion.

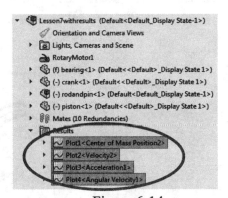

Figure 6-14

Make sure you have completed a simulation before proceeding to the interference check. Right click the root assembly *Lesson6* in the *MotionManager* tree and then select *Check Interference*. The *Find Interferences Over Time* dialog box appears (Figure 6-15).

Pick *poston<1>* and *rod<1>* (expanding *roadandpin<1>*) from the *MotionManager* tree to be included in the interference check. The *Start Frame*, *End Frame*, and *Increment* allow you to specify the motion frame used as the starting position, final position, and increment in between for the interference check. We will enter *1*, *201*, and *1*, for the *Start Frame*, *End Frame*, and *Increment*, respectively. Click the *Find Now* button (circled in Figure 6-15) to start the interference check.

After pressing the *Find Now* button, the mechanism starts moving, and the crank rotates a complete cycle. At the same time, the list in the *Find Interferences Over Time* dialog box continuously grows. The list at the lower half of the dialog box shows all interference conditions detected. The frame, simulation time, parts that caused the interference, and the volume of the interference detected are listed.

Select the *Index* number of an entity (for example, *5*, as shown in Figure 6-16) to enable the *Details* button and the *Magnifying* button at the lower right corner. Selecting *Details* displays more information about the interference. Click the *Magnifying* button to zoom in to the location of the interference. As shown in Figure 6-17, the interference occurs between the inner face of the piston and the end face of the rod. You may want to choose the *Front* view to see the interference (Figure 6-18).

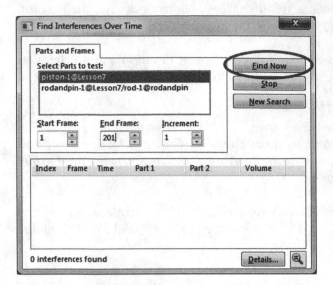

Figure 6-15  The *Find Interference Over Time* Dialog Box

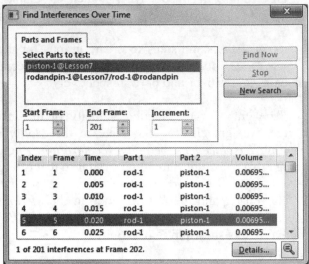

Figure 6-16  Interference Check Results

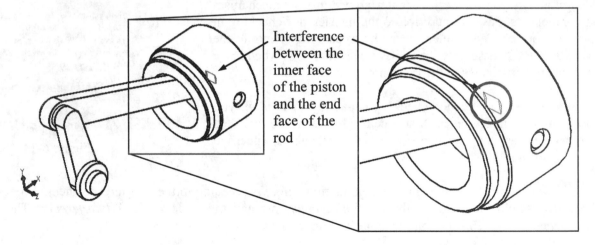

Figure 6-17  Interference Identified

Note that you may see a different result than those of Figures 6-16 to 6-18 if you choose subassembly *roadandpin<1>* instead of *rod<1>*. The interference checking capability of *Motion 2017 SP1.0* is not able to capture interference correctly for subassemblies due to most likely a software bug. Therefore, make sure you expand the subassembly *roadandpin<1>* and then choose *rod<1>* part.

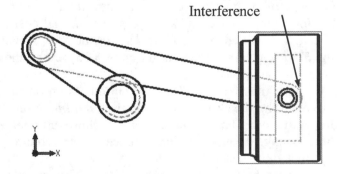

Figure 6-18  Interference (Front View)

A workaround of this problem is to use the interference detection capability in the assembly mode as mentioned earlier. You may adjust the view of the motion model (for example, like that of Figure 6-18) and choose *Tools > Evaluate > Interference Detection* to find the interference between parts and subassemblies for the entire mechanism.

Close the dialog box and save the model.

### Creating and Running a Dynamic Analysis

Duplicate the motion model and rename it *Dynamic*.

A force simulating the engine firing load (acting along the negative *X*-direction) will be added to the piston for a dynamic simulation. It will be more realistic if the force can be applied when the piston starts moving to the left (negative *X*-direction) and can be applied only for a selected short period of time.

In order to do so, we will have to define measures that monitor the position of the piston for the firing load to be activated. Unfortunately, such a capability is not available in *Motion 2017*. Therefore, the force is simplified as a step function of *3* lb$_f$ along the negative *X*-direction and applied for *0.1* seconds. The force will be defined as a point force at the center point of the end face of the piston, as shown in Figure 6-19.

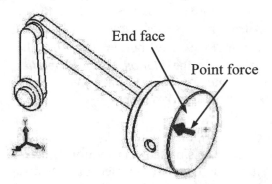

Figure 6-19 Pick the End Face for Force Application

Before adding the force, we will turn off the rotary motor. From the *MotionManager* tree, right click *RotaryMotor1*, and choose *Suppress*. The motor will be turned off. Note that if you run a simulation now, nothing will happen since there is no motion driver or force defined (gravity has not been added).

Now we are ready to add the force. Click the *Force* button ↖ from the *Motion* toolbar. In the *Force* dialog box (Figure 6-20), choose *Linear Force* (default) for *Force Type* and *Action Only* (default) for *Direction*, and then pick the end face of the piston, as shown in Figure 6-19 for the location of the applied force.

The *Face<1>@piston-1* is now listed in the dialog box (Figure 6-20), and a force arrow will appear on the end face of the piston in the graphics area, indicating that the force will be applied to the center of the end face and in the direction that is normal to the face; i.e., in the positive *X*-direction. Click the small arrow button ↗ to flip the force direction. The force arrow in the graphics area should now be reversed like that of Figure 6-19. That is, the force will be applied to the center of the end face of the piston and in the negative *X*-direction.

Scroll down the box to define *Force Function*. Click the *Function* tab, choose *Step* function, and enter the following:

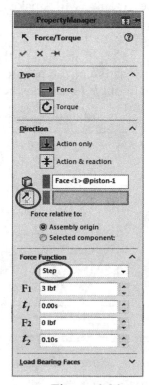

Figure 6-20

*Initial Value F1: 3*
*Start Time t1: 0*
*Final Value F2: 0*
*End Time t2: 0.1*

Click checkmark ✔ on top of the dialog box to accept the definition. A force node (*Force1*) should appear in the *MotionManager* tree.

There is no capability to graph the external force in *2017 SP1.0*. One approach is to graph the force as a reaction force by choosing *Result and Plots* 📈 from the *Motion* toolbar after a successful motion analysis, as shown in Figure 6-21. In fact, the force varies from *3* at *0* second to *0* at *0.1* seconds. During the simulation, the force is activated at the beginning; i.e., *0* second.

Before running a simulation, we will increase the number of frames in order to see more refined results and graphs. Click the *Motion Study Properties* button ⚙ from the *Motion* toolbar. In the *Motion Study Properties* dialog box, enter *500* for *Frames per second*, and click the checkmark on top.

Click *Calculate* button 📊 from the *Motion* toolbar to simulate the motion. A one-second simulation will be carried out. After a short moment, you should see the mechanism starts moving. The mechanism will move and the crank will make several turns before reaching the end of the simulation duration.

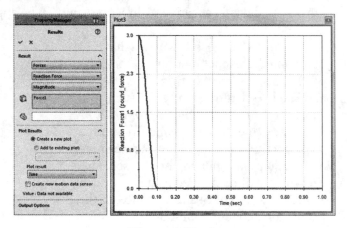

Figure 6-21

Graphs created in the kinematic simulation are available for displaying dynamic simulation results, such as the piston position, etc. As shown in Figure 6-22, the piston moves along the negative *X*-direction for about *0.15* seconds before reversing its direction.

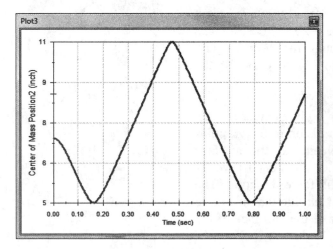

Figure 6-22  *X*-Position of the Piston

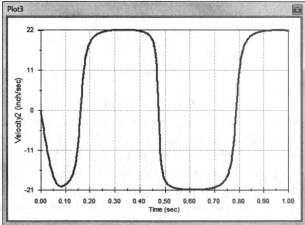

Figure 6-23  *X*-Velocity of the Piston

It is also evident in the velocity graph shown in Figure 6-23 that the velocity changes signs at three instances (close to *0.15*, *0.48*, and *0.78* seconds). Recall that the force was applied for the first *0.1* seconds. Had the force application lasted longer, the piston could be continuously pushed to the left (negative *X*-direction) even when the piston reaches the left end and tries to move to the right (due to inertia). As a result, the crank would have been oscillating at the left of the center of the bearing; i.e., between *0* and *180* degrees about the *Z*-axis, without making a complete turn.

Next, we will graph the reaction force on the piston due to the applied force. The reaction force can be graphed only at assembly mates that constrain certain degrees of freedom. As discussed in *Lesson 1*, these reaction forces at mates (or joints) are keeping the moving bodies connected following Newton's 3$^{rd}$ Law. Graphing the reaction force in *Motion 2017* is a little tricky due to the redundant mates assigned by the *Motion* solver internally. When graphing the reaction forces, you may see zero forces or moments if redundant degrees of freedom of mates are selected. If this happens, you may want to pick another mate that constrains the same degree of freedom to graph the non-zero reaction forces.

In this example, we will define a plot by selecting the piston and then picking *Concentric3* that connects the pin to the piston to graph the reaction force.

From the *MotionManager* tree, right-click *piston<1>* and choose *Create Motion Plot*. In the *Results* dialog box, choose *Forces*, select *Reaction Force*, and *X Component*. Then, clear the face listed and pick *Concentric3* (from *MotionManager* tree), as shown in Figure 6-24. Click the checkmark to accept the definition.

A warning message box appears, indicating that the motion model has redundant constraint which can lead to invalid results, and asking you to replace redundant constraints with bushings. Select *No* to move on. A graph like that of Figure 6-25 should appear. There are three peaks in Figure 6-25 representing the time when the largest reaction forces occur at the joint, which occurs at close to *0.15*, *0.48*, and *0.78* seconds; i.e., when the piston reverses its moving direction. The results seem to make sense. Note that in order to capture the peak force, you will need to increase *Frames per Second* for motion analysis. In this example, 500 frames per second is adequate.

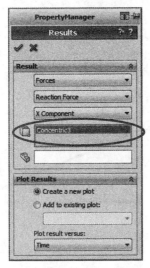

Figure 6-24

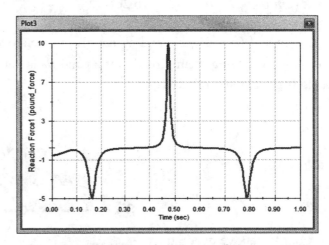

Figure 6-25  *X*-Reaction Force of *Concentric3*

Save the model. We will carry out theoretical calculations to verify the simulation results next. We will focus on kinematic analysis.

### 6.4  Result Verifications

In this section, we verify the motion analysis results using kinematic analysis theory often found in mechanism design textbooks. Note that in kinematic analysis, position, velocity, and acceleration of given points or axes in the mechanism are analyzed.

In kinematic analysis, forces and torques are not involved. All bodies (or links) are assumed massless. Hence, mass properties defined for bodies are not influencing the analysis results.

The slider-crank mechanism is a planar kinematic analysis problem. A vector plot that represents the positions of joints of the planar mechanism is shown in Figure 6-26. The vector plot serves as the first step in computing position, velocity, and accelerations of the mechanism.

The position equations of the system can be described by the following vector summation,

$$\mathbf{Z}_1 + \mathbf{Z}_2 = \mathbf{Z}_3 \tag{6.1}$$

where

$$\mathbf{Z}_1 = Z_1 \cos \theta_A + i Z_1 \sin \theta_A = Z_1 e^{i\theta_A}$$
$$\mathbf{Z}_2 = Z_2 \cos \theta_B + i Z_2 \sin \theta_B = Z_2 e^{i\theta_B}$$
$$\mathbf{Z}_3 = Z_3, \text{ since } \theta_C \text{ is always } 0.$$

The real and imaginary parts of Eq. 6.1, corresponding to the $X$ and $Y$ components of the vectors, can be written as

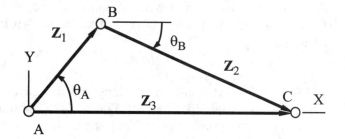

$$Z_1 \cos \theta_A + Z_2 \cos \theta_B = Z_3 \tag{6.2a}$$
$$Z_1 \sin \theta_A + Z_2 \sin \theta_B = 0 \tag{6.2b}$$

Figure 6-26  Vector Plot of the
Slider-Crank Mechanism

In Eqs. 6.2a and 6.2b, $Z_1$, $Z_2$, and $\theta_A$ are given. We are solving for $Z_3$ and $\theta_B$. Equations 6.2a and 6.2b are non-linear. Solving them directly for $Z_3$ and $\theta_B$ is not straightforward. Instead, we will calculate $Z_3$ first, using trigonometric relations; i.e.,

$$Z_2^2 = Z_1^2 + Z_3^2 - 2Z_1 Z_3 \cos \theta_A$$

Hence,

$$Z_3^2 - 2Z_1 \cos \theta_A Z_3 + Z_1^2 - Z_2^2 = 0$$

Solving $Z_3$ from the above quadratic equation, we have

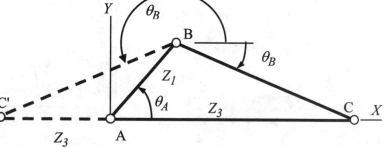

Figure 6-27  Two Possible Configurations

$$Z_3 = \frac{2Z_1 \cos \theta_A \pm \sqrt{(2Z_1 \cos \theta_A)^2 - 4(Z_1^2 - Z_2^2)}}{2} \tag{6.3}$$

where two solutions of $Z_3$ represent the two possible configurations of the mechanism shown in Figure 6-27. Note that point C can be either at C or C' for any given $Z_1$ and $\theta_A$.

From Eq. 6.2b, $\theta_B$ can be solved by

$$\theta_B = sin^{-1}\left(\frac{-Z_1 \sin\theta_A}{Z_2}\right) \tag{6.4}$$

Similarly, $\theta_B$ has two possible solutions corresponding to vector $\mathbf{Z}_3$.

Taking derivatives of Eqs. 6.2a and 6.2b with respect to time, we have

$$-Z_1 \sin\theta_A \dot\theta_A - Z_2 \sin\theta_B \dot\theta_B = \dot Z_3 \tag{6.5a}$$

$$Z_1 \cos\theta_A \dot\theta_A + Z_2 \cos\theta_B \dot\theta_B = 0 \tag{6.5b}$$

where $\dot\theta_A = \dfrac{d\theta_A}{dt} = \omega_A$ is the angular velocity of the rotary motor, which is a constant. Note that Eqs. 6.5a and 6.5b are linear functions of $\dot Z_3$ and $\dot\theta_B$. Rewrite the equations in a matrix from

$$\begin{bmatrix} Z_2 \sin\theta_B & 1 \\ Z_2 \cos\theta_B & 0 \end{bmatrix}\begin{bmatrix} \dot\theta_B \\ \dot Z_3 \end{bmatrix} = \begin{bmatrix} -Z_1 \sin\theta_A \dot\theta_A \\ -Z_1 \cos\theta_A \dot\theta_A \end{bmatrix} \tag{6.6}$$

Equation 6.6 can be solved by

$$\begin{aligned}
\begin{bmatrix} \dot\theta_B \\ \dot Z_3 \end{bmatrix} &= \begin{bmatrix} Z_2 \sin\theta_B & 1 \\ Z_2 \cos\theta_B & 0 \end{bmatrix}^{-1}\begin{bmatrix} -Z_1 \sin\theta_A \dot\theta_A \\ -Z_1 \cos\theta_A \dot\theta_A \end{bmatrix} \\[2mm]
&= \frac{1}{-Z_2 \cos\theta_B}\begin{bmatrix} 0 & -1 \\ -Z_2 \cos\theta_B & Z_2 \sin\theta_B \end{bmatrix}\begin{bmatrix} -Z_1 \sin\theta_A \dot\theta_A \\ -Z_1 \cos\theta_A \dot\theta_A \end{bmatrix} \\[2mm]
&= \frac{1}{-Z_2 \cos\theta_B}\begin{bmatrix} Z_1 \cos\theta_A \dot\theta_A \\ Z_1 Z_2 \cos\theta_B \sin\theta_A \dot\theta_A - Z_1 Z_2 \sin\theta_B \cos\theta_A \dot\theta_A \end{bmatrix} \\[2mm]
&= \begin{bmatrix} -\dfrac{Z_1 \cos\theta_A \dot\theta_A}{Z_2 \cos\theta_B} \\[4mm] -\dfrac{Z_1\left(\cos\theta_B \sin\theta_A \dot\theta_A - \sin\theta_B \cos\theta_A \dot\theta_A\right)}{\cos\theta_B} \end{bmatrix}
\end{aligned} \tag{6.7}$$

Hence

$$\dot{\theta}_B = -\frac{Z_1 \cos\theta_A \, \dot{\theta}_A}{Z_2 \cos\theta_B} \tag{6.8}$$

and

$$\dot{Z}_3 = Z_1\left( \tan\theta_B \cos\theta_A \, \dot{\theta}_A - \sin\theta_A \, \dot{\theta}_A \right) \tag{6.9}$$

In this example, $Z_1 = 3$, $Z_2 = 8$, and the initial conditions are $\theta_A(0) = \pi/2$ and $\theta_B(0) = \sin^{-1}(3/8)$. Note that these initial conditions are different from those of *Motion* model, where $\theta_A(0) = 0$. You may graph the angular displacement of the crank (see Figure 6-28). The angular displacement starts at zero (when the crank is in the upright position; that is $\theta_A(0) = \pi/2$), increases to *180* degrees (crank rotates 180 degrees, vertically downward), drop to *–180* degrees (with the same physical configuration), and comes back to zero. The discontinuous angular displacement shown in *Motion* is artificial, resulting from the way that the angular displacement was measured in *Motion*.

The solutions can be implemented using a spreadsheet. The *Excel* spreadsheet file, *lesson6.xls*, can be found at the publisher's web site. As shown in Figure 6-29, Columns A to I represent time, $Z_1$, $Z_2$, $\dot{\theta}_A$, $\theta_A$, $Z_3$, $\theta_B$, $\dot{Z}_3$, and $\dot{\theta}_B$, respectively. Note that in this calculation, $Z_3(0)>0$ is assumed, hence $\theta_B(0)<0$ (clockwise), as illustrated in Figure 6-27. This is consistent with the initial configuration we created in the *Motion* model.

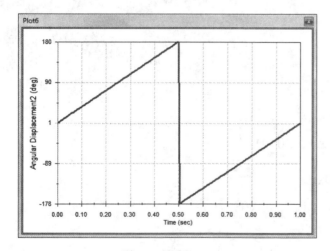

Figure 6-28

Figure 6-29  The *Excel* Spreadsheet

| Time | Z1 | Z2 | ThetaADot | ThetaA | Z3 | ThetaB | Z3Dot | ThetaBDot |
|---|---|---|---|---|---|---|---|---|
| 0 | 3 | 8 | 6.2831853 | 1.5707963 | 7.4161985 | -0.38439677 | -18.84955592 | -8.92075E-15 |
| 0.005 | 3 | 8 | 6.2831853 | 1.6022123 | 7.3225649 | -0.38419718 | -18.60088462 | 4.573894613 |
| 0.01 | 3 | 8 | 6.2831853 | 1.6336282 | 7.2302189 | -0.38359868 | -18.33468044 | 9.141065151 |
| 0.015 | 3 | 8 | 6.2831853 | 1.6650441 | 7.1392455 | -0.38260215 | -18.05202481 | 13.6948191 |
| 0.02 | 3 | 8 | 6.2831853 | 1.69646 | 7.0497242 | -0.38120906 | -17.75400499 | 18.22852667 |
| 0.025 | 3 | 8 | 6.2831853 | 1.727876 | 6.9617292 | -0.37942145 | -17.44170793 | 22.73565112 |
| 0.03 | 3 | 8 | 6.2831853 | 1.7592919 | 6.8753291 | -0.37724191 | -17.11621437 | 27.20977791 |
| 0.035 | 3 | 8 | 6.2831853 | 1.7907078 | 6.7905873 | -0.37467359 | -16.7785931 | 31.64464224 |
| 0.04 | 3 | 8 | 6.2831853 | 1.8221237 | 6.7075617 | -0.3717202 | -16.4298956 | 36.0341548 |
| 0.045 | 3 | 8 | 6.2831853 | 1.8535397 | 6.6263051 | -0.36838594 | -16.07115099 | 40.37242526 |
| 0.05 | 3 | 8 | 6.2831853 | 1.8849556 | 6.5468652 | -0.36467554 | -15.70336136 | 44.65378343 |
| 0.055 | 3 | 8 | 6.2831853 | 1.9163715 | 6.4692849 | -0.36059421 | -15.32749755 | 48.87279776 |
| 0.06 | 3 | 8 | 6.2831853 | 1.9477874 | 6.3936021 | -0.3561476 | -14.94449538 | 53.02429123 |
| 0.065 | 3 | 8 | 6.2831853 | 1.9792034 | 6.3198504 | -0.35134184 | -14.55525229 | 57.10335428 |
| 0.07 | 3 | 8 | 6.2831853 | 2.0106193 | 6.2480586 | -0.34618344 | -14.16062455 | 61.10535506 |
| 0.075 | 3 | 8 | 6.2831853 | 2.0420352 | 6.1782517 | -0.34067932 | -13.76142486 | 65.0259467 |
| 0.08 | 3 | 8 | 6.2831853 | 2.0734512 | 6.1104507 | -0.33483677 | -13.3584205 | 68.86107194 |
| 0.085 | 3 | 8 | 6.2831853 | 2.1048671 | 6.0446727 | -0.3286634 | -12.95233183 | 72.60696501 |
| 0.09 | 3 | 8 | 6.2831853 | 2.136283 | 5.9809314 | -0.32216717 | -12.54383139 | 76.260151 |
| 0.095 | 3 | 8 | 6.2831853 | 2.1676989 | 5.9192373 | -0.31535629 | -12.13354324 | 79.81744286 |
| 0.1 | 3 | 8 | 6.2831853 | 2.1991149 | 5.859598 | -0.30823928 | -11.72204283 | 83.27593626 |
| 0.105 | 3 | 8 | 6.2831853 | 2.2305308 | 5.802018 | -0.30082486 | -11.30985719 | 86.63300236 |
| 0.11 | 3 | 8 | 6.2831853 | 2.2619467 | 5.7464997 | -0.293122 | -10.89746537 | 89.88627898 |
| 0.115 | 3 | 8 | 6.2831853 | 2.2933626 | 5.693043 | -0.28513984 | -10.48529931 | 93.0336601 |
| 0.12 | 3 | 8 | 6.2831853 | 2.3247786 | 5.6416457 | -0.27688771 | -10.07374487 | 96.07328418 |
| 0.125 | 3 | 8 | 6.2831853 | 2.3561945 | 5.592304 | -0.26837509 | -9.663143114 | 99.0035213 |
| 0.13 | 3 | 8 | 6.2831853 | 2.3876104 | 5.5450122 | -0.25961158 | -9.253791793 | 101.8229595 |
| 0.135 | 3 | 8 | 6.2831853 | 2.4190263 | 5.4997635 | -0.2506069 | -8.845946968 | 104.5303905 |
| 0.14 | 3 | 8 | 6.2831853 | 2.4504423 | 5.4565499 | -0.24137089 | -8.439824772 | 107.124795 |
| 0.145 | 3 | 8 | 6.2831853 | 2.4818582 | 5.4153621 | -0.23191343 | -8.035603273 | 109.6053274 |
| 0.15 | 3 | 8 | 6.2831853 | 2.5132741 | 5.3761904 | -0.2222445 | -7.633424397 | 111.9713009 |
| 0.155 | 3 | 8 | 6.2831853 | 2.54469 | 5.3390243 | -0.21237411 | -7.233395904 | 114.2221727 |
| 0.16 | 3 | 8 | 6.2831853 | 2.576106 | 5.3038528 | -0.20231234 | -6.835593393 | 116.3575285 |
| 0.165 | 3 | 8 | 6.2831853 | 2.6075219 | 5.2706646 | -0.19206928 | -6.440062308 | 118.3770685 |
| 0.17 | 3 | 8 | 6.2831853 | 2.6389378 | 5.2394483 | -0.18165504 | -6.046819934 | 120.2805933 |
| 0.175 | 3 | 8 | 6.2831853 | 2.6703538 | 5.2101926 | -0.17107975 | -5.655857375 | 122.0679898 |
| 0.18 | 3 | 8 | 6.2831853 | 2.7017697 | 5.182886 | -0.16035355 | -5.267141493 | 123.7392188 |

Figures 6-30 and 6-31 show the graphs of data in Columns F and H, respectively. Comparing Figures 6-30 and 6-31 with Figures 6-10 and 6-11, the simulation analysis results are verified.

Note that the accelerations of a given joint in the mechanism can be formulated by taking one more derivative of Eqs. 6.5a and 6.5b with respect to time. The resulting two coupled equations can be solved using *Excel* spreadsheet. This is left as an exercise.

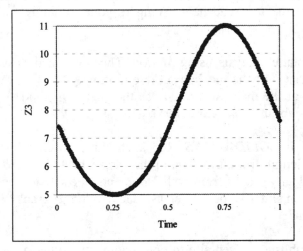

Figure 6-30  Position of the Piston (Column F)

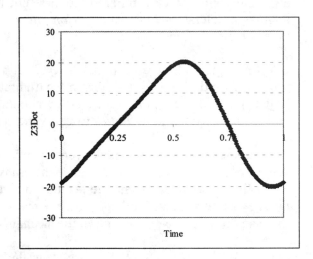

Figure 6-31  Velocity of the Piston (Column H)

**Exercises:**

1.  Derive the acceleration equations for the slider-crank mechanism by taking derivatives of Eqs. 6.5a and 6.5b with respect to time. Solve these equations for the linear acceleration of the piston and the angular acceleration of the mate *Concentric2*, using a spreadsheet. Compare your solutions with those obtained from *Motion*.

2.  Use the same slider-crank model to conduct a static analysis using *Motion*. The static analysis in *Motion* should give you equilibrium configuration(s) of the mechanism due to gravity (turn on the gravity). Show the equilibrium configuration(s) of the mechanism and use the energy method you learned from Sophomore *Statics* to verify the equilibrium configuration(s).

3.  Change the length of the crank from *3* to *5* in. in *SOLIDWORKS*. Repeat the kinematic analysis discussed in this lesson. Then, change the crank length from *3* to *5* in. in the spreadsheet (*Microsoft Excel* file, *lesson6.xls*). Generate position and velocity graphs from both *Motion* and the spreadsheet. Do they agree with each other? Does the maximum slider velocity increase due to a longer crank? Is there any interference occurring in the mechanism?

4.  Download five *SOLIDWORKS* parts from the publisher's web site to your computer (folder name: *Exercise 6-4*).

    (i)   Use these five parts, i.e., bearing, crankshaft, connecting rod, piston pin, and piston (see Figure E6-1), to create an assembly like the one shown in Figure E6-2. Note that the crankshaft must orient at *45°* CCW, as shown in Figure E6-2.

    (ii)  Create a motion model for kinematic analysis. Conduct motion analysis by defining a driver that drives the crankshaft at a constant angular speed of *1,000* rpm.

    (iii) Use the spreadsheet *lesson6.xls* to calculate the piston velocity. Compare your calculations with those obtained from *Motion*.

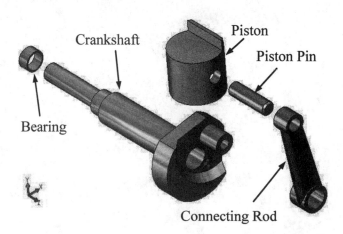

Figure E6-1  Five *SOLIDWORKS* Parts

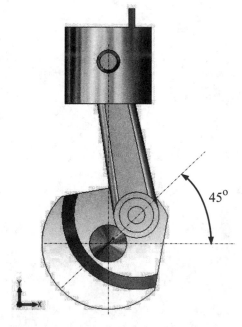

Figure E6-2  Assembled Configuration

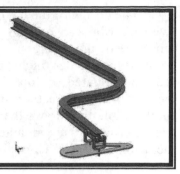

# Lesson 7: A Rail Carriage Example

## 7.1 Overview of the Lesson

In this lesson, we will learn an advanced feature in *SOLIDWORKS 2017*, the *Path Mate*, which was first added to *SOLIDWORKS 2008*. Most assembly mates impose constraints to part motion between regular surfaces, such as flat surfaces and cylindrical surfaces. As a result, mating parts are allowed to translate or rotate along a fixed direction.

In path mate, a vertex in a part is moving along a curve (or a loop composed of several curves) of the mating part. As a result, path mate allows a part to move along a curve slot, a groove, or fluting, varying its moving direction specified by the path curve. In addition, the pitch, yaw and roll of the moving part can be defined to resemble the physical conditions. Such a capability offered in *Motion 2017* supports animation and motion analysis for a different set of applications involving curvilinear motion.

In this lesson, we will use the rail and the top portion of the carriage of a bathroom transporting device shown in Figure 7-1 as an example. This was an undergraduate design project carried out a few years ago.

This transporting device was created for the purpose of transporting a disabled woman from wheelchair to toilet and to shower bench without human assistance. It also transports the person from toilet or shower bench back to the wheelchair. The device is compact (to fit into a very small bathroom), durable, and tailored to help the person to overcome her physical disability. The design features a three-button remote control that moves the person to the toilet, shower, and back to the wheelchair; a scissor lift with a linear actuator that provides lift; a carriage on a rail system that carries the person to designated locations; and a body support that safely holds the person while the system transports the person to designated locations. A second actuator mounted on top of the scissor provides a 90-degree rotation to the body support when the carriage is moved to the toilet so that the user will be properly oriented on top of the toilet. And a motor and a cable system are employed to pull the carriage.

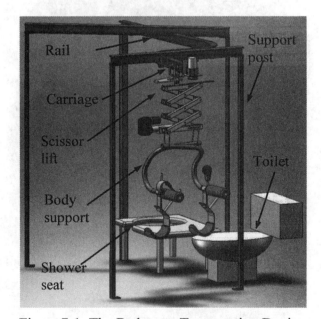

Figure 7-1 The Bathroom Transporting Device

The design of the device was extremely challenging since it has to accommodate a severely disabled woman, who can only use her right hand to operate the device. She will pull her wheelchair to the entrance of the bathroom, right in front of the device. She will use her right hand to move the two leg supports under her thighs, place the two arm supports under her arms, and press a button on the remote control mounted on top of the right arm support. The button pressing will trigger the actuator of the scissor lift to contract, creating a lift to move her out of the wheelchair. Then, a motor will be activated to pull a cable that draws the carriage along the curve rail and transport her to the toilet or shower bench. Position sensors are mounted on top of the rail to detect the location of the carriage and activate motor or actuators for desired motions.

While designing the device, path mates were employed to assemble the carriage to the rail, allowing the carriage to move along the rail. This motion animation supports the verification of the design concept, facilitates communications within the design team, and offers a demonstration of the design to the project sponsors and users before physically building it.

## 7.2  The Rail Carriage Example

The rail and carriage are important features of the device. The rail is a curve I-beam, created by sweeping an I-cross section sketch along an open loop curve composed of three straight lines and two circular arcs, as shown in Figure 7-2. The carriage consists of a base plate, two steerers, and four wheels, as shown in Figure 7-3. The wheels are sitting on the top faces of the bottom flange of the rail. A cable connecting to a motor is pulling the steerers to move the carriage along the rail. A universal joint under the base plate connects the body support.

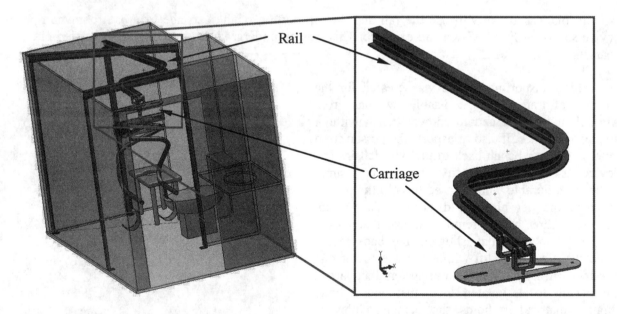

Figure 7-2  The Rail and Carriage Subsystems

### *SOLIDWORKS Parts and Assembly*

Parts and assembly have been created in *SOLIDWORKS*. There are four parts and one subassembly: *Rail.SLDPRT*, *Base Plate.SLDPRT*, *Steerer.SLDPRT*, *Wheels.SLDPRT*, and *Carriage.SLDASM*. There are also two assembly files, *Lesson7.SLDASM* and *Lesson7withresults.SLDASM*, created for this example.

In this lesson, we will start with *Lesson7.SLDASM*, where standard mates were employed. You will see the undesired motion of the carriage in this model and the deficiency of the assembly mates employed. We will then replace the standard mates with path mates to create desired carriage motion. The assembly with path mates has been created in *Lesson7withresults.SLDASM*. The assembly file with correct results can be found in the subfolder, *Lesson 7 Saved from 2009 Version*. You may want to open *Lesson7withresults.SLDASM* under this subfolder to see the desired motion of the carriage.

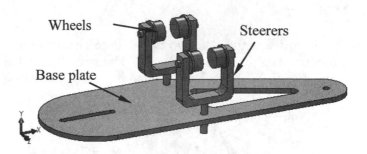

Figure 7-3  The Carriage Subassembly

The assembly *Lesson7.SLDASM* consists of one part and one subassembly: the rail (*Rail.SLDPRT*) and the carriage (*Carriage.SLDASM*). The rail is fixed to the ground, and the carriage is fully assembled to the rail by using five standard mates: three coincident and two tangent mates, as shown in Figure 7-4. You may want to expand the *Mates* node in the browser to see the list of mates. Move the mouse pointer to any of the mates; you should see the geometric entities selected for the mate highlighted in the graphics area.

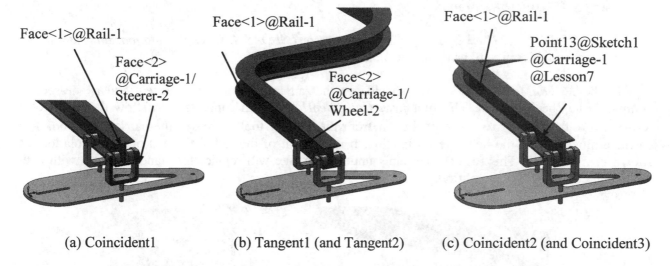

(a) Coincident1          (b) Tangent1 (and Tangent2)          (c) Coincident2 (and Coincident3)

Figure 7-4  Assembly Mates Defined for the Rail Carriage Example

The first mate (*Coincident1*) shown in Figure 7-4a aligns the two end faces of the rail and the steerer. This mate defines the initial position of the carriage, which will be suppressed to allow the carriage to move. The next two are tangent mates shown in Figure 7-4b, which constrain the outer cylindrical surface of the wheel (the two on the left side of the rail) to be tangent to the top face of the bottom flange of the rail. The final two are coincident mates, which place the mid-points of the axes in the respective steerers on the left surface of the rail web (see Figure 7-4c). In the carriage assembly, there are two sketches (*Sketch1* and *Sketch2*), where the two mid-points are defined (please see Figure 7-5). The carriage is fully assembled to the rail. You will not be able to move the carriage by dragging it in the graphics area.

*Motion Model*

In this example, the carriage will be given an initial velocity of 25 in./second along the negative Z-direction, as shown in Figure 7-6. The overall simulation is one second. A motion model of standard mates discussed above can be found in *Lesson7.SLDASM*.

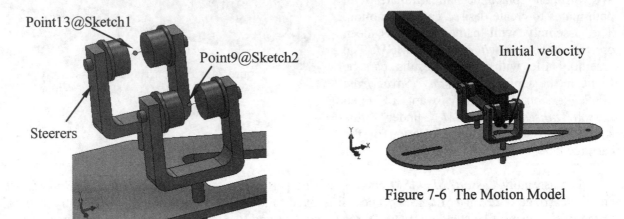

Figure 7-6  The Motion Model

Figure 7-5  Mid-Points of axes in the Carriage

## 7.3  Using *SOLIDWORKS Motion*

Start *SOLIDWORKS* and open assembly file *Lesson7.SLDASM*. Click the *Motion Study* tab at the bottom of the graphics area to bring up the *MotionManager* window.

In the *MotionManager* tree, you should see that an initial velocity is listed, and a one-second simulation has been created. The first mate *Coincident1* has been suppressed to allow the carriage to move. Click the *Play from start* ▮▶ button from the *Motion* toolbar to see the carriage motion. The carriage moves along the negative Z-direction. It moves out of the rail at the end of the simulation, as shown in Figure 7-8. This result is certainly undesirable. We will replace the standard mates with path mates for a desired carriage motion.

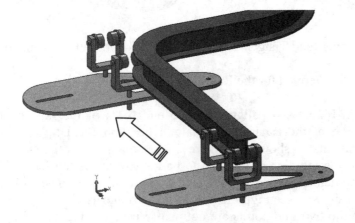

Figure 7-7  Carriage Moved Out of the Rail

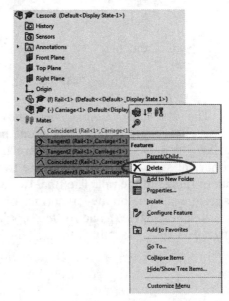

Figure 7-8

*Defining Path Mates*

First, we delete all mates except for *Coincident1*.

Click the *Model* tab to go back to *SOLIDWORKS* assembly. From the assembly browser, expand the *Mates* node. Click *Tangent1*, press the *Shift* key, and then click *Coincident3*. All four mates are highlighted like that of Figure 7-8. Right click and choose *Delete* to delete all four mates. Click *Yes* button in the *Confirm Delete* dialog box four times (or choose *Yes to All*).

We will create two path mates. The first one will be defined by selecting *Point13* in the carriage and the sweep path of the rail (an open loop created in *Sketch2* of the rail part under feature *Sweep1*). Turn on the point display to see the points by choosing *View > Hide/Show > Points* from the pull-down menu. Do the same for sketch (*View > Hide/Show > Sketches*).

Go back to *SOLIDWORKS* by clicking the *Model* tab at the bottom of the graphics area.

From *SOLIDWORKS* browser (see Figure 7-9), expand *Rail <1>*, then expand *Sweep1*, and choose *Sketch2*. You should see the sweep path highlighted in the graphics area (Figure 7-10a). This is the path we will pick for the path mate. Expand *Carriage<1>* and choose *Sketch1*. *Point13* will appear in the graphics area, as shown in Figure 7-10b. This is the vertex we will pick for the first path mate. Similarly, we will pick *Point9* of *Sketch2* to create the second path mate.

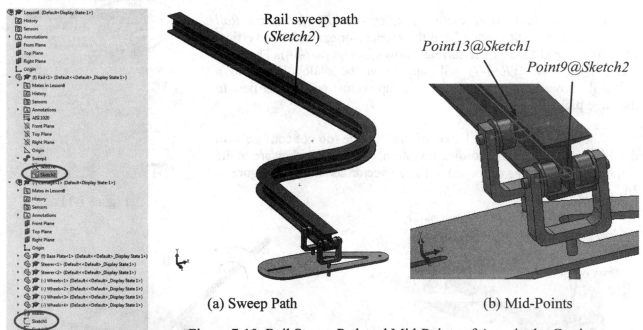

Rail sweep path
(*Sketch2*)

*Point13@Sketch1*

*Point9@Sketch2*

(a) Sweep Path                                                    (b) Mid-Points

Figure 7-9

Figure 7-10  Rail Sweep Path and Mid-Points of Axes in the Carriage

Insert a mate by choosing *Insert > Mate* from the pull-down menu or clicking the *Mate* button 🔩 from the assembly toolbar. The *Mate* dialog box will appear (Figure 7-11). Choose *Advanced Mates* and click *Path Mate*. The *Component Vertex* of the *Mate Selections* is active and ready for you to pick a geometric entity. From the graphics area, pick *Point13*. Next, the *Path Selection* is active and ready for you to pick. We will first click *Selection Manager* button (right under the *Path Selection*), click *Select Open Loop* button in a set of buttons appear in the graphics area, as shown in Figure 7-12, and then pick the beginning of the sweep path. Right click to accept the open loop chosen. Both entities should now be highlighted in the graphics area and listed in *Mate Selections*. Click the checkmark ✔ to accept the path mate.

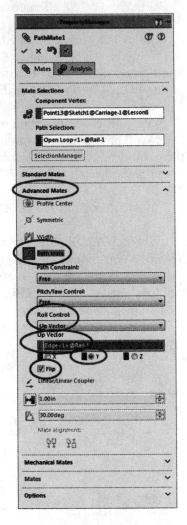

Figure 7-11

For the second path mate, we will pick *Point9* and the same sweep path (make sure you choose open loop). Click the checkmark ✔ to accept the second mate. Click the checkmark one more time to close the dialog box. In the assembly browser, two path mates are added (*PathMate1* and *PathMate2*).

Pick the carriage in the graphics area and drag it. You should be able to drag the carriage through the path. However, the subassembly rolls when you move the carriage, which is not desirable.

Next, we will control the rolling motion of the carriage (see Figure 7-13 for pitch, yaw, and roll). We will create the *Roll Control* using the *Path Mate* dialog box (Figure 7-11), where we will pick a vertical edge for the up vector, then select *Y* axis to align with the vector in order to constrain the rolling motion.

From the browser, right click *PathMate1*, and choose *Edit Feature* 🔧.

From the *PathMate1* dialog box, choose *Up Vector* for *Roll Control*, pick a vertical edge from the model (for example the vertical edge of the end face of the rail, as shown in Figure 7-14). The edge picked (*Edge<1>@Rail-1*) will appear in the dialog box, and a vertically downward arrow (that is, the up vector) will appear next to the edge picked.

Choose *Y* to align the *Y*-axis of the carriage (do not confuse with the *Y*-axis of the global coordinate system, even though they are in the same direction in this example) to the up vector, as shown in Figure 7-14.

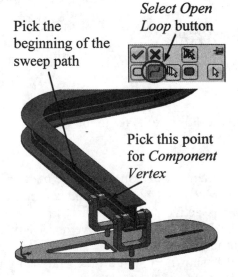

Pick the beginning of the sweep path

*Select Open Loop* button

Pick this point for *Component Vertex*

Figure 7-12  Pick Entities for the Path Mate

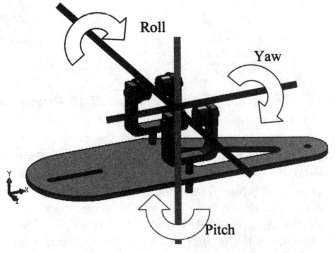

Roll

Yaw

Pitch

Figure 7-13  Roll, Yaw, and Pitch of the Carriage

If the carriage is positioned on the top of the rail, click *Flip* to bring it to the right position; i.e., below the rail.

Click the checkmark ✔ on top of the dialog box to accept the modified mate definition. We will leave all the path, pitch, yaw, and roll control of the second path mate to *Free* since they have been constrained in the first path mate.

Now, pick the carriage in the graphics area and drag it. You should be able to drag the carriage through the path without rolling.

Unsuppress *Coincident1* mate to bring the carriage back to its initial position. Suppress it to allow motion.

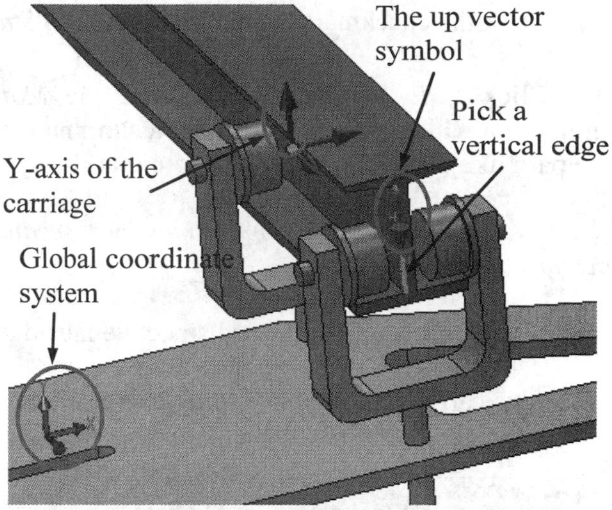

Save the model.

We are ready to conduct a motion simulation.

**Figure 7-14  Up Vector and Component *Y*-Axis**

### *Defining and Running Simulation*

Click the *Motion Study* tab at the bottom of the graphics area. In the *MotionManager* window, choose *Motion Analysis* for the study.

First, drag the end time key to the four-second mark (to allow the carriage to go through the entire path). We will refine the analysis time steps and choose a different solver (we will use *WSTIFF*).

Click the *Motion Study Properties* button ⚙ from the *Motion* toolbar. In the *Motion Study Properties* dialog box, enter *500* for *Frame per second*, and click *Advanced Options* button (see Figure 7-15).

In the *Advanced Motion Analysis Options* dialog box (Figure 7-16), choose *WSTIFF*, and then click *OK*.

Note that *WSTIFF* and the default solver *GSTIFF* are similar in formulation and behavior. Both use a backwards difference formulation. They differ in that the *GSTIFF* coefficients are calculated assuming a constant step size, whereas *WSTIFF* coefficients are a function of the step size. If the step size changes suddenly during integration, *GSTIFF* introduces a small error, while *WSTIFF* can handle step size changes without loss of accuracy. Sudden step size changes occur whenever there are discontinuous forces, discontinuous motions or abrupt events, such as the carriage turning in this example.

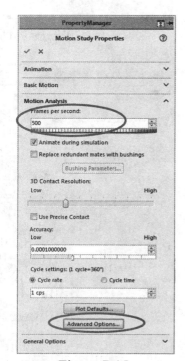

Figure 7-15

Click the checkmark on top of the *Motion Study Properties* dialog box to accept the definition.

Click *Calculate* button  from the *Motion* toolbar to simulate the motion. A four-second simulation will be carried out. After a short moment, you should see that the carriage starts moving along the path like that of Figure 7-17. Note that if you see a warning dialog box, stating that

*The playback speed or frames per second settings for this motion study will result in poor performance ...*

Simply click *No* to proceed since we intend to use high frame rate for a more accurate solution.

This carriage motion is desirable. Be sure to save the model before exiting from *SOLIDWORKS*.

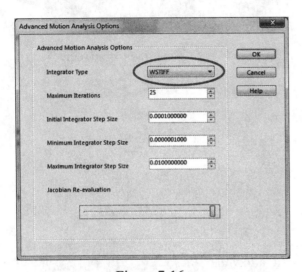

Figure 7-16

Figure 7-17  Desired Carriage Motion

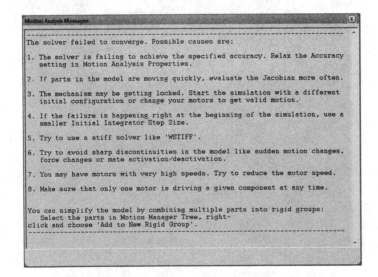

Figure 7-18

Note that you may encounter a solution error (see Figure 7-18) while you are running a motion analysis, if you are using *SOLIDWORKS 2017 SP1.0*. The error message appears when the carriage makes the first turn. Unfortunately, there is nothing the user can do to fix the problem since it is a software bug recognized by *SOLIDWORKS*. One possible workaround (less desirable) is to suppress the second path mate in the motion study and use solid body contacts to keep the second steer in contact with the track. A "workaround" motion model can be found in the Lesson 7 folder, called *Lesson 7 Solid Contact Workaround*.

Also, as mentioned earlier, a motion model with desired carriage motion saved from version 2009 (without the "workaround") can be found in the subfolder *Lesson 7 Saved from 2009 Version*. You may open the assembly file, *Lesson7withresults.SLDASM*, and click *Play from Start* button ▐► (do not click the *Calculate* button ▦ ) to review the carriage motion.

**Exercises:**

1. Create a motion model for the wheel to go through the curve groove of the ground shown in Figure E7-1. Note that the short axle of the wheel must align to the two small cuts of the groove horizontally, as shown in Figure E7-2.

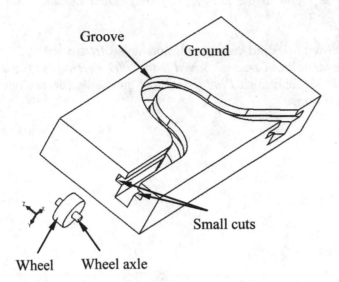

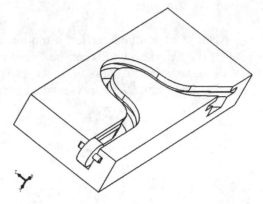

Figure E7-2  Wheel Properly Aligned

Figure E7-1  The Wheel and Ground Parts

# Lesson 8: A Compound Spur Gear Train

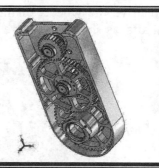

## 8.1  Overview of the Lesson

In this lesson we will learn how to simulate motion for a spur gear train system. A gear train is a set or system of gears arranged to transfer torque or energy from one part of a mechanical system to another. A gear train consists of driving gears that are mounted on the input shaft, driven gears mounted on the output shaft, and idler gears that interpose between the driving and driven gears in order to maintain the direction of the output shaft to be the same as the input shaft or to increase the distance between the drive and driven gears. There are different kinds of gear trains, such as a simple gear train, compound gear train, epicyclic gear train, etc., depending on how the gears are shaped and arranged as well as the functions they intend to perform. The gear train we are simulating in this lesson is a compound gear train system, in which two or more gears are used to transmit torque or energy. All gears included in this lesson are spur gears; therefore, the shafts that these gears mounted on are in parallel.

In *Motion*, gear pair is defined as a *Mechanical Mate*, which couples two rotational motions of respective gear axes. Neither *SOLIDWORKS* nor *Motion* cares about the detailed geometry of the gear pair; i.e., if the gear teeth mesh adequately. You may even use cylinders or disks to represent the gears. No detailed tooth profile is necessary for any of the computations involved in the gear mate. Apparently, force and moment between a pair of teeth in contact will not be calculated in gear train simulations. However, there are other important data being calculated by *Motion*, such as reaction force exerting on the driven shaft (for a dynamic analysis), which is critical for mechanism design. In any case, pitch circle diameters are essential for defining gear pair and gear trains in *Motion*. Gear ratio of the gear train, which is defined by the ratio of the angular velocities of the output and input gears, is determined by the pitch circle diameters of the individual gear pairs in the gear train system.

Although cylinders or disks are sufficient to model gears in *SOLIDWORKS*, we will use more realistic gear models in this lesson. All gears in the example are shown with detailed geometry, including teeth. In addition, detailed parts, including shafts, bearings, screws and aligning pins are included for a realistic gear train system, as shown in Figure 8-1. In this gear train simulation, we will focus more on graphical animation, less on computations of physical quantities. We will add a rotary motor to drive the input shaft.

## 8.2  The Gear Train Example

### Physical Model

The gear train example we are using for this lesson is part of a gearbox designed for an experimental lunar rover. The gear train is located in a gear box which is part of the transmission system of the rover, driven by a motor powered by solar energy. The purpose of the gear train is to convert a high-speed rotation and small torque generated by the motor to a low speed rotation and large torque output in order to drive the wheels of the rover. The gear train consists of four spur gears mounted on three parallel shafts, as shown in Figure 8-1.

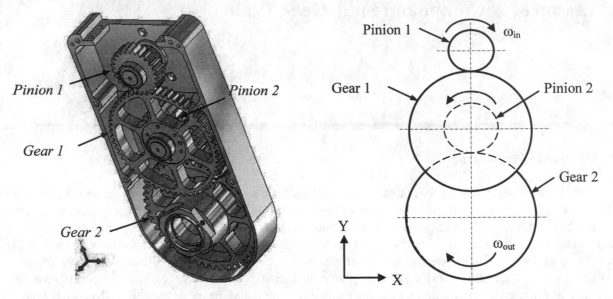

Figure 8-1  The Gear Train System in Rover          Figure 8-2  Schematic View of the Gear Train

The four spur gears form two gear pairs: *Pinion 1* and *Gear 1*, and *Pinion 2* and *Gear 2*, as depicted in Figure 8-1 and 8-2. Note that *Pinion 1* is the driving gear that connects to the motion driver; e.g., a rotary motor. The motor rotates in a clockwise direction, therefore driving *Pinion 1*. *Gear 1* is the driven gear of the first gear pair, which is mounted on the same shaft as *Pinion 2*. Both rotate in a counterclockwise direction. *Gear 2* is driven by *Pinion 2*, and rotates in a clockwise direction. Note that the diameters of the pitch circles of the four gears are 50, 120, 60, and 125 mm, respectively; and the numbers of teeth are 25, 60, 24, and 50, respectively. Therefore, the circular pitch $P_c$, the diametral pitch $P_d$, and module $m$ of the first gear pair are, respectively

$$p_c = \frac{\pi d_{p1}}{N_{p1}} = \frac{\pi d_{g1}}{N_{g1}} = \frac{\pi(50)}{25} = \frac{\pi(120)}{60} = 6.283 \text{ mm}, \quad p_d = \frac{N_{p1}}{d_{p1}} = \frac{N_{g1}}{d_{g1}} = \frac{25}{50} = \frac{60}{120} = 0.5 \text{ mm}^{-1},$$

$$m = \frac{d_{p1}}{N_{p1}} = \frac{d_{g1}}{N_{g1}} = \frac{50}{25} = \frac{120}{60} = 2 \text{ mm}.$$

(8.1)

For the second gear pair, we have

$$p_c = \frac{\pi d_{p2}}{N_{p2}} = \frac{\pi d_{g2}}{N_{g2}} = \frac{\pi(60)}{24} = \frac{\pi(125)}{50} = 7.854 \text{ mm}, \quad p_d = \frac{N_{p2}}{d_{p2}} = \frac{N_{g2}}{d_{g2}} = \frac{24}{60} = \frac{50}{125} = 0.4 \text{ mm}^{-1},$$

$$m = \frac{d_{p1}}{N_{p1}} = \frac{d_{g1}}{N_{g1}} = \frac{50}{24} = \frac{125}{50} = 2.5 \text{ mm}.$$

(8.2)

Therefore, the gear ratio $g_r$ of the gear train is:

$$g_r = \frac{\omega_{out}}{\omega_{in}} = \frac{N_{p1}}{N_{g1}}\frac{N_{p2}}{N_{g2}} = \frac{25}{60}\frac{12}{50} = \frac{1}{5}, or$$

$$= \frac{d_{p1}}{d_{g1}}\frac{d_{p2}}{d_{g2}} = \frac{50}{120}\frac{60}{125} = \frac{1}{5}$$

(8.3)

where $\omega_{out}$ and $\omega_{in}$ are the output and input angular velocities of the gear train system, respectively; $d_{p1}$, $d_{g1}$, $d_{p2}$, and $d_{g2}$ are the pitch diameters of the respective four gears; and $N_{p1}$, $N_{g1}$, $N_{p2}$, and $N_{g2}$ are the number of teeth of the four respective gears. The gear ratio of the gear train shown in Figure 1 is *1:5*; i.e., the angular velocity is reduced *5* times at the output shaft. Theoretically, the torque output will increase *5* times if there is no loss due to, e.g., friction. Note that we will use *MMGS* units system for this lesson.

### SOLIDWORKS Parts and Assembly

In this lesson, *SOLIDWORKS* parts of the gear train have been created. There are six model files: *gbox_housing.SLDPRT*, *gbox_input.SLDPRT*, *gbox_middle.SLDPRT*, *gbox_output.SLDPRT*, *Lesson8.SLDASM*, and *Lesson8withresults.SLDASM*. We will start with *Lesson8.SLDASM*, in which the gears are fully assembled to the housing. In addition, the assembly file *Lesson8withresults.SLDASM* consists of a complete simulation model with simulation results. You may want to open *Lesson8withresults.SLDASM* to preview the gear motion.

Note that the housing part in *SOLIDWORKS* was converted directly from *Pro/ENGINEER* part. The three gear parts in *SOLIDWORKS* were converted from respective *Pro/ENGINEER* assemblies. There were nine, nine, and six distinct parts within the three gear assemblies, respectively. These three *Pro/ENGINEER* assemblies (and associated parts) were first converted to *SOLIDWORKS* as assemblies. Parts in each assembly were then merged into a single gear part in *SOLIDWORKS*. The detailed part and assembly conversions as well as merging multiple parts into a single part in *SOLIDWORKS* can be found in Appendix C.

In the *SOLIDWORKS* assembly model *Lesson8.SLDASM* (and *Lesson8withresults.SLDASM*), there are nine assembly mates. The first three mates, *Concentric1*, *Coincident1*, and *Coincident2*, assemble the input gear to the housing. The input gear is fully constrained. Note that before entering *Motion*, we will suppress *Coincident2* in order to allow the input gear to rotate along the Z-axis (see Figures 8-3a, b, and c). Similarly, the next three mates, *Concentric2*, *Coincident3* and *Coincident4*, assemble the middle gear to the housing. Again, we will suppress *Coincident4* to allow the middle gear to rotate about the Z-axis (see Figures 8-3d, e and f). The third set of mates, *Concentric3*, *Coincident5*, and *Coincident6*, does the same for the output gear (see Figures 8-3g, h and i). Similarly, *Coincident6* will be suppressed to allow for rotation. Note that the three suppressed mates are created to properly orient the three gears so that the gear teeth mesh properly between pairs. The three mates to be suppressed also define the initial configuration of the gear train.

As mentioned earlier, one important factor for the animation to "look right" is to mesh the gear teeth properly. You may want to use the *Front* view and zoom in to the tooth mesh areas to check if the two pairs of gears mesh properly (see Figure 8-4). They should mesh well, which is accomplished by the three coincident mates that will be suppressed before entering *Motion*. Note that the tooth profile is represented by straight lines, instead of more popular ones such as involutes, just for simplicity.

If you turn on the axis display (*View > Hide/Show > Axes*), axes parallel to the Z-axis that pass through the center of the respective gears will appear (please see Figure 8-5). These axes are necessary for creating the gear mates.

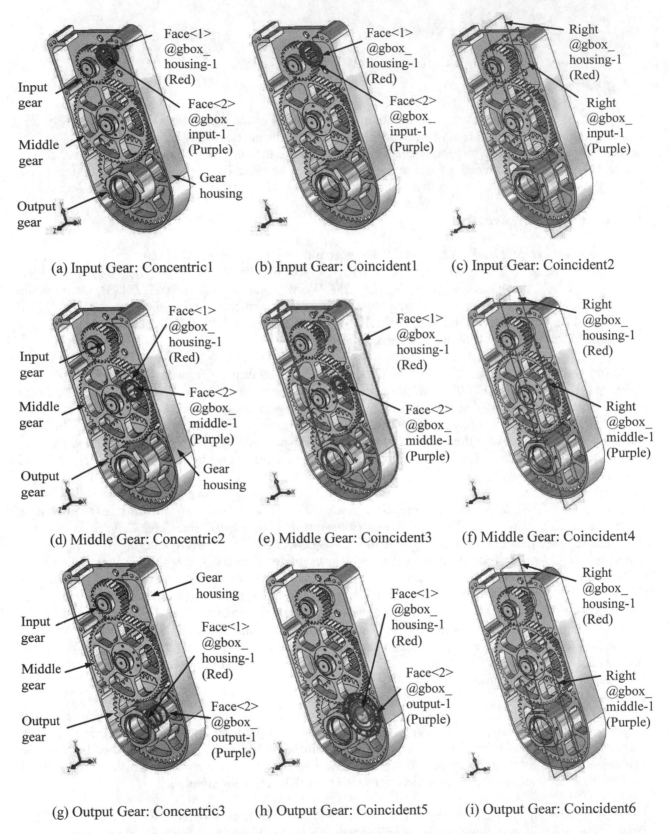

(a) Input Gear: Concentric1      (b) Input Gear: Coincident1      (c) Input Gear: Coincident2

(d) Middle Gear: Concentric2     (e) Middle Gear: Coincident3     (f) Middle Gear: Coincident4

(g) Output Gear: Concentric3     (h) Output Gear: Coincident5     (i) Output Gear: Coincident6

Figure 8-3  Assembly Mates Defined in *Lesson8.SLDASM*

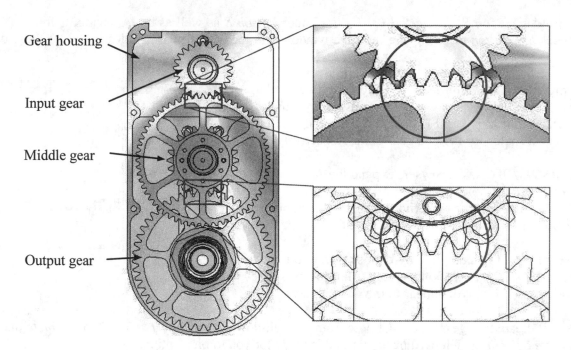

Figure 8-4  Gear Teeth Properly Meshed

### Simulation Model

The gear housing will be defined as the ground part. All three gears will rotate with respect to their respective axes. The four gears will be meshed into two gear pairs: *Pinion 1* with *Gear 1*, and *Pinion 2* with *Gear 2*, as discussed earlier. In *SOLIDWORKS*, gear pairs are created by selecting two axes of the respective gears (or cylinders) using *Mechanical Mates* option. Note again that before the gear mates can be created, we will suppress the three coincident mates that orient the gears; i.e., *Coincident2*, *Coincident4* and *Coincident6*, in order to allow desired gear rotation motion. The axis of the input gear will be driven by a rotary motor with a constant angular velocity of 360 degrees/sec. We will conduct a kinematic analysis for this example.

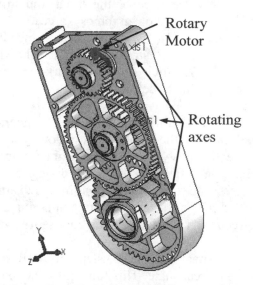

Figure 8-5  Gear Train Motion Model

### 8.3  Using *SOLIDWORKS Motion*

Start *SOLIDWORKS* and open *Lesson8.SLDASM*.

Note that when you open the assembly, you will see a message window, as shown in Figure 8-6, indicating that *SOLIDWORKS* is unable to locate *gbox_input.sldasm*. *SOLIDWORKS* is trying to locate the assembly from which the input gear part was created. Since the input gear assembly and its associated parts were translated from *Pro/ENGINEER* and are not available in the *Lesson 8* folder, *SOLIDWORKS* is unable to locate it. It is fine to choose *Suppress components* and not to locate *gbox_input.sldasm*. Not locating the assembly file for the input gear part will not affect the motion simulation in this lesson. After that, *SOLIDWORKS* will ask you to locate the middle gear assembly and output gear assembly. Choose *Suppress components* for both. The assembly files that *SOLIDWORKS* is looking for are actually located in the subfolder *Lesson 8_Converted ProE Files* under *Lesson 8* as well as the Appendix C folder. You may choose *Browse for file* from the message window and locate the missing files in one of these two folders.

Before entering *Motion* there are two things we need to do. First, we will suppress three assembly mates, *Coincident2*, *Coincident4* and *Coincident6*. Second, we will create two gear mates for the two gear pairs, respectively.

From *SOLIDWORKS* browser, expand the *Mates* node, click *Coincident2*, and choose *Suppress*. The mate *Coincident2* will become inactive. Repeat the same to suppress *Coincident4* and *Coincident6*. Save the model.

Next, turn on the axis view by choosing from the pull-down menu *View > Hide / Show > Axes*. All three axes, one for each gear assembly, will appear in the graphics area.

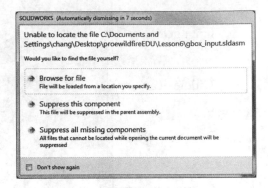

Figure 8-6

We will create two gear mates. Choose from the pull-down menu *Insert > Mate*. In the *Mate* dialog box, the *Mate Selections* field will be active, and is ready for you to pick entities.

Pick the axes of the input and middle gears from the graphics area. Choose *Mechanical Mates*, click *Gear*, and enter *50mm* and *120mm* for *Ratio*, as shown in Figure 8-7. Click the checkmark on top to accept the mate definition.

Note that if the axes of the two gears are pointing in the opposite direction, you will have to click *Reverse* box (right below the *Ratio* text field in the *Mate* dialog box) to correct the rotation direction. In this example, all three axes are pointing in the same direction. Therefore, do not choose *Reverse*.

Repeat the same steps to define the second gear mate. This time, pick the axes of the middle and output gears, and enter *60mm* and *125mm* for *Ratio*. Click the checkmark to accept the mate, and click it one more time to close the dialog box. Two new mates, *GearMate1(gbox_input<1>, gbox_middle<1>)* and *GearMate2(gbox_middle<1>,gbox_output<1 >)*, are now listed under *Mates*.

Figure 8-7  Defining Gear Mate

Now we are ready to enter *Motion*.

Click the *Motion Study* tab at the bottom of the graphics area to bring up the *MotionManager* window. We will add a rotary motor and create a one-second simulation.

### Adding a Rotary Motor

Click the *Motor* button 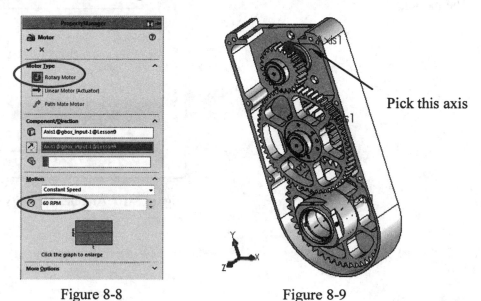 from the *Motion* toolbar to bring up the *Motor* dialog box (Figure 8-8). Choose *Rotary Motor* (default). Move the pointer to the graphics area, and pick *Axis1* of the input gear, as shown in Figure 8-9. A circular arrow appears indicating the rotational direction of the rotary motor. A clockwise direction (along Z-axis) is desired. You may change the direction by clicking the direction button right under *Component/Direction,* if necessary. Choose *Constant speed* and enter *60* RPM for speed. Click the checkmark on top of the dialog box to accept the motor definition. You should see a *RotaryMotor1* added to the *MotionManager* tree.

Pick this axis

Figure 8-8                                                    Figure 8-9

We are ready to run a simulation. If you are just interested in gear motion animation, you may simply choose *Animation* option for the study.

### Running Simulation

Choose *Animation* for the study, and drag the end time key to one-second mark. Click the *Motion Study Properties* button from the *Motion* toolbar. In the *Motion Study Properties* dialog box, enter *60* for *Frames per second*, and click the checkmark on top. Click the *Calculate* button from the *Motion* toolbar to simulate the motion. A one second simulation will be carried out. After a short moment, you should see the gears start moving. The input gear rotates *360* degrees as expected, which drives all the other gears to rotate.

You may want to adjust the playback speed to 10% (0.1x) to slow down the animation since the total simulation duration is only one second. Look closer to the gear teeth. Both gear pairs should mesh correctly. The animation looks good.

### Saving and Reviewing Results

We will graph the angular velocity of the output gear. In order to show result graphs, we will have to choose *Motion Analysis* for the study, and rerun the simulation. Click the *Motion Study Properties* button

, enter *60* for *Frames per second*, and click the checkmark on top. Click the *Calculate* button from the *Motion* toolbar to simulate the motion.

Right click *gbox_output<1>* in the *MotionManager* tree, and choose *Create Motion Plot*. In the *Results* dialog box, choose *Displacement/Velocity/Acceleration*, select *Angular Velocity*, and then *Z Component*. The graph will appear similar to that of Figure 8-10, which shows that the output velocity is a constant of *72* degrees/sec. Note that this magnitude is one fifth of the input velocity since the gear ratio is *1:5*. Both the input (*Pinion 1*) and output gears (*Gear 2*) rotate in the same direction. *Motion* gives good results.

Save the model.

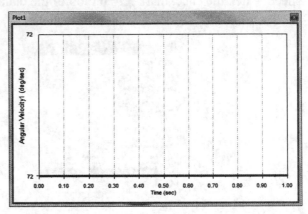

Figure 8-10  Angular Velocity of Output Gear

**Exercises:**

1.  The same gear train will be used for this exercise. Create a constant torque for the input gear (*gbox_input.SLDPRT*) about the Z-axis. Turn on friction for all three axles (*Steel-Dry/Steel-Dry*). Define and run a *2*-second dynamic simulation for the gear train.

    (i)   What is the minimum torque that is required to rotate the input gear, and therefore, the entire gear train?

    (ii)  If the torque applied to the input gear is *100* mm N, what is the output angular velocity of the gear train at the end of the *2*-second simulation? Verify the simulation result using your own calculation.

    (iii) Create a graph for the reaction moment between gears of the first gear pair (*GearMate1*) due to the *100* mm N torque. What is the reaction moment obtained from simulation?

**Notes:**

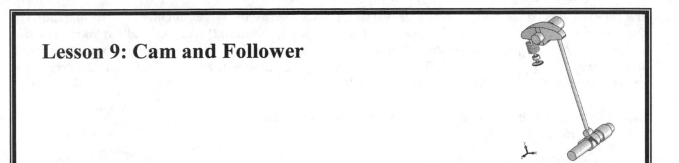

# Lesson 9: Cam and Follower

## 9.1   Overview of the Lesson

In this lesson, we will learn cam and follower (or cam-follower). Physically, a cam-follower is a device that converts rotary motion into linear motion. In *Motion*, a cam and follower can be assembled using a cam-follower assembly mate. A simple form of a cam is a rotating disc with a variable radius so that its profile is not circular but oval or egg-shaped. When the disc rotates, its edge (or side face) pushes against a follower, which may be a small wheel at the end of a lever or the end face of the lever or rod itself. The follower will thus rise and fall at exactly the same amount as the variation in radius of the cam. By profiling a cam appropriately, a desired cyclic pattern of straight-line motion, in terms of position, velocity, and acceleration, can be produced.

We will learn to create a motion model and simulate the control of opening and closing of an inlet or exhaustive valve, usually found in internal combustion engines, using cam-follower connections. In a design such as that of Figure 9-1, the drive for the camshaft is taken from the crankshaft through a timing chain, which keeps the cams synchronized with the movement of the piston so that the valves open and close at a precise instant. The mechanism we will be working with consists of bushings, camshaft, pushrod, rocker, valve, valve guide and spring, as shown in Figure 9-1. The cam-follower connects the camshaft and the pushrod. When the cam on the camshaft pushes the pushrod up, the rocker rotates and pushes the valve on the other side downward. The spring surrounding the valve gets compressed and opens up the inlet for air to flow into the combustion chamber.

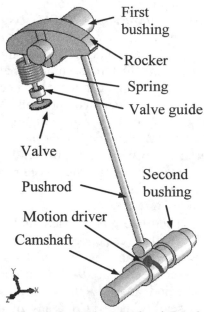

Figure 9-1  The Mechanism of Engine Inlet or Outlet Valve

## 9.2   The Cam and Follower Example

***Physical Model***

The camshaft and the rocker will rotate about the axes of their respective revolute joints connecting them to their respective bearings (defined as ground body). The camshaft is driven by a motor of constant velocity of *600* rpm (or *10* rev/sec). The profile of the cam consists of two circular arcs of *0.25* and *0.5* in. radii, respectively, as shown in Figure 9-2. The lower arc is concentric with the shaft, and the center of the upper arc is *0.52* in. above the center of the shaft. When the camshaft rotates, the cam mounted on the shaft pushes the pushrod up by up to *0.27* in. (that is, *0.52+0.25−0.5 = 0.27*). As a result, the rocker will rotate and push the valve at the other end downward at a frequency of *10* times/sec. The valve will move again up to *0.27* in. downward since the pushrod and the valve are positioned at an equal distance from the rotation axis of the rocker. When the camshaft rotates where the larger circular arc (*0.5* in. radius) of the cam is in contact with the follower (in this example, the pushrod), the pushrod has room to move downward. At this point, the rocker will rotate back since the spring is being uncompressed. As a result, the valve will move up, and therefore, close the inlet. The valve will be open for about *120* degree per cycle, based on the cam design shown in Figure 9-2.

The units system chosen for this example is *IPS* and all parts are made up of steel.

### *SOLIDWORKS Parts and Assembly*

The cam-follower system consists of seven parts, bushing (two), valve guide, camshaft, pushrod, rocker, and valve, as shown in Figure 9-1. In addition, there are two assembly files, *Lesson9.SLDASM* and *Lesson9withresults.SLDASM*, that you may download from the publisher's web site.

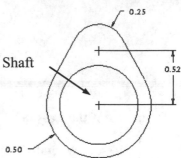

Figure 9-2  The Cam Profile

We will start with *Lesson9.SLDASM*, in which the parts are adequately assembled. In this assembly the first bushing is anchored (ground) and the second bushing and the valve guide are fully constrained. These three parts will be fixed (ground). The remaining four parts will be movable.

Same as before, the assembly file *Lesson9withresults.SLDASM* contains a complete simulation model with simulation results. You may want to open the assembly to see the motion animation of the mechanism. In the assembly where a motion model is completely defined, a mate has been suppressed to allow movement between components. You can also see how the parts move by right clicking in the graphics area, choosing *Move Component*, and dragging any movable parts; for example the camshaft to rotate with respect to the second bushing. The whole mechanism should move accordingly.

There are eighteen assembly mates, including four coincident, three concentric, seven distance, two parallel, one tangent, and one cam-mate-tangent, as listed in the browser (see Figure 9-3). You may want to expand the *Mates* node in *SOLIDWORKS* browser to review the list of assembly mates. Click any of the mates; you should see the entities selected to define the assembly mate highlighted in the graphics screen.

As mentioned earlier, the first bushing, *bushing<1>*, shown in the browser is anchored to the assembly. The second bushing (*bushing<2>*) and the valve guide (*valve guide<1>*) were fully assembled to *bushing<1>*. The first three mates, *Coincident1*, *Distance1*, and *Distance2*, were employed to assemble *bushing<2>* to *bushing<1>*. And the next three distance mates assemble the valve guide to *bushing<1>*.

Figure 9-3  Assembly Mates Listed in the Browser

Note that the distance mates are essentially coincident mate with distance between entities. The distance mates were created to properly position the second bushing with respect to the first bushing. All three parts will be regarded as the ground part in *Motion*.

The next two mates, *Concentric1* and *Coincident2*, assemble the rocker (*rocker<1>*) to the first bushing (*bushing<1>*), allowing a rotation degree of freedom about the *Z*-axis, as shown in Figures 9-4a and b. *Motion* will map a revolute joint between the rocker and the first bushing.

The next part assembled is the pushrod (*pushrod<1>*). The pushrod was assembled to the rocker using *Tangent1*, *Distance6*, and *Coincident3* mates, as shown in Figures 9-4c, d, and e, respectively. As a result, the pushrod is allowed to move vertically at a distance of *1.25* in. from the *Right* plane of the first bushing (*Distance6*), at the same time maintaining tangency between the top of the cylindrical surface of the pushrod and the socket surface of the rocker.

The next part is the camshaft (*cam_shaft<1>*). The camshaft was first assembled to the second bushing (*bushing<2>*) and then to the pushrod. *Concentric2* aligns the camshaft and the second bushing. *Coincident4* mates the center plane of the camshaft to that of pushrod, as shown in Figures 9-4f and g. As a result, the camshaft is allowed to rotate about the *Z*-axis.

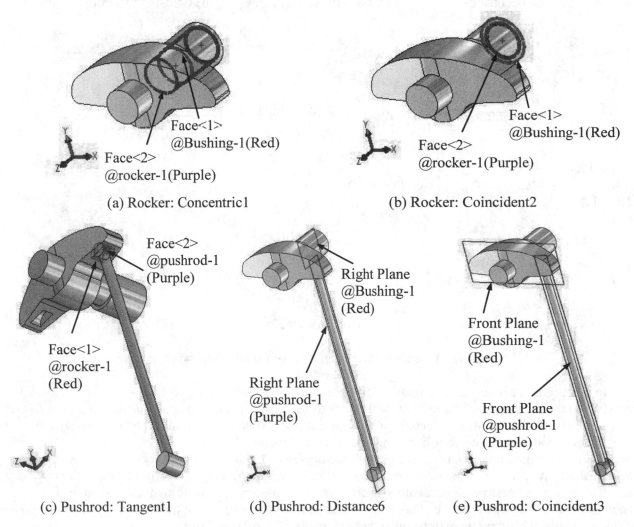

(a) Rocker: Concentric1          (b) Rocker: Coincident2

(c) Pushrod: Tangent1          (d) Pushrod: Distance6          (e) Pushrod: Coincident3

Figure 9-4  Assembly Mates Defined for the Mechanism

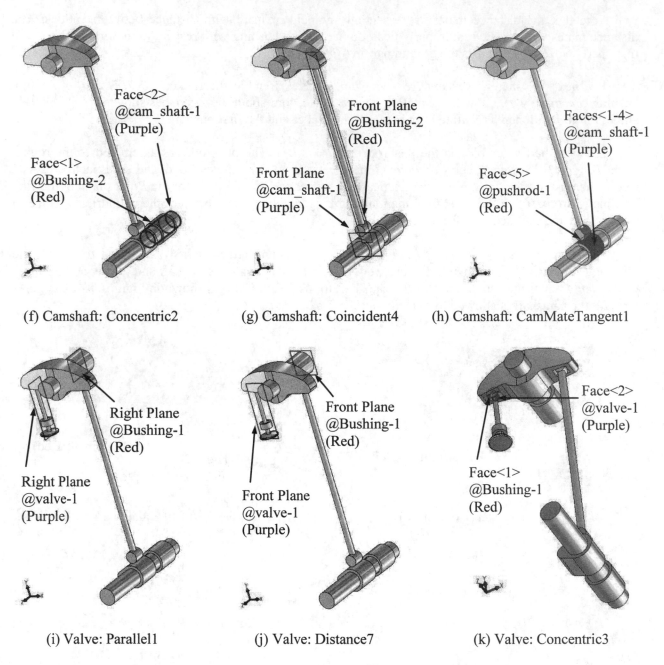

(f) Camshaft: Concentric2    (g) Camshaft: Coincident4    (h) Camshaft: CamMateTangent1

(i) Valve: Parallel1    (j) Valve: Distance7    (k) Valve: Concentric3

Figure 9-4  Assembly Mates Defined for the Mechanism (Cont'd)

In addition, the *CamMateTangent1* defines a cam and follower between the cam surface of the camshaft and the cylindrical surface at the bottom of the pushrod. Note that all surrounding surfaces on the cam and follower must be selected for the cam-follower joint. In this case, four surfaces are selected for the cam (mounted on the camshaft) and one surface is included on the follower, as shown in Figure 9-4h. Note that the assembly should have overall one degree of freedom at this point. You may either rotate the camshaft, the rocker, or move the pushrod vertically to see the relative motion of the assembly. Use the *Undo* button to restore the assembly to its original configuration. Note that the last mate *Parallel2* listed in the browser was inactive. This mate was defined to properly orient the camshaft to the *Top* plane of the assembly. This mate is mainly for setting up an initial condition for simulation.

The next and the final part assembled is the valve. The valve was assembled to the first bushing using *Parallel1* and *Distance7* mates, which restrict the valve to move vertically on the *X-Y* plane, as shown in Figures 9-4i and j. The third mate, *Concentric3*, aligns the top cylindrical surface of the valve with the other socket surface of the rocker, as shown in Figure 9-4k. Note that the valve will move vertically along the *Y*-direction, and slightly deviate from the nominal distance of *1.25* in. between its *Right* plane and the *Right* plane of the first bushing. This slight movement is necessary due to the rotation of the rocker. That is why the mate *Parallel1* (between the *Right* plane of the valve and the *Right* plane of the first bushing) is employed instead of a distance mate.

Note that this set of assembly mates employed is not the only way to assemble the individual parts for a valid motion model. You may try different combinations to mate these parts. As long as the assembly is physically valid, *Motion* is usually able to simulate the motion of a valid assembly that is physically meaningful.

### Simulation Model

Once we suppress the mate *Parallel2(cam_shaft<1>,Top Plane)*, the mechanism has one free degree of freedom. Any rotation on the shaft will be sufficient to uniquely determine the position, velocity, and acceleration of any parts in the mechanism. Physically the total degree of freedom of the system is *1*.

The motion model is shown in Figure 9-5, where the *Z*-rotation of the camshaft is driven by a rotary motor with a constant angular velocity of 3,600 degrees/sec; i.e., 600 rpm, about the *Z*-axis of the global coordinate system. In addition, a spring surrounding the valve will be created in order to provide a vertical force to push the rocker up, and therefore, close the valve. The spring has a spring constant of *10* $lb_f$/in and an unstretched length of *1.25* in. The spring is created between the bottom face of the rocker and top face of the valve guide.

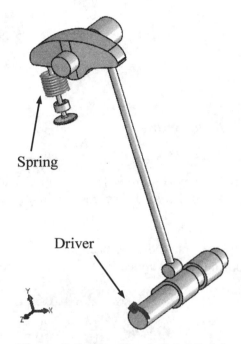

Spring

Driver

Figure 9-5  The Simulation Model

### 9.3   Using *SOLIDWORKS Motion*

Start *SOLIDWORKS* and open assembly file *Lesson9.SLDASM*. Again, always check the units system. Make sure that *IPS* units system is chosen for this example. Suppress mate *Parallel2*.

Click the *Motion Study* tab at the bottom of the graphics area to bring up the *MotionManager* window.

### Adding a Rotary Motor

Click the *Motor* button ![motor icon] from the *Motion* toolbar to bring up the *Motor* dialog box (Figure 9-6). Choose *Rotary Motor* (default). Move the pointer to the graphics area, and pick a circular edge of a cam, as shown in Figure 9-7. A circular arrow appears indicating the rotational direction of the rotary motor. A counterclockwise direction (Z-direction) is desired. Choose *Constant speed* and enter *600* RPM (10 revolutions per second) for speed. Click the checkmark ![checkmark] on top of the dialog box to accept the motor definition. You should see a *RotaryMotor1* added to the *MotionManager* tree.

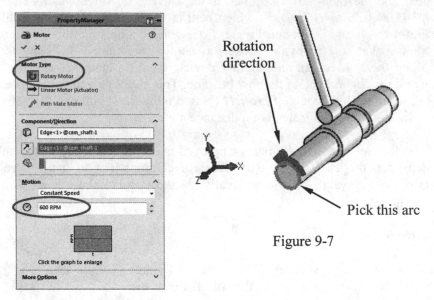

Figure 9-7

Figure 9-6

### Defining Spring

Click the *Spring* button 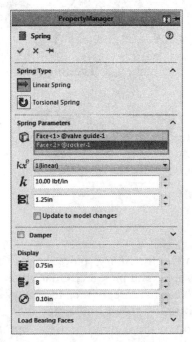 from the *Motion* toolbar to bring up the *Spring* dialog box. In the *Spring* dialog box (Figure 9-8), the empty field right underneath the *Spring parameters* label is active for you to pick entities to define ends of the spring. Rotate the view and pick the bottom face of the rocker (see Figure 9-9).

Rotate the view back, and then pick the top face of the valve guide, as shown in Figure 9-9. Also, a spring should appear in the graphics area, connecting the center points of the two faces.

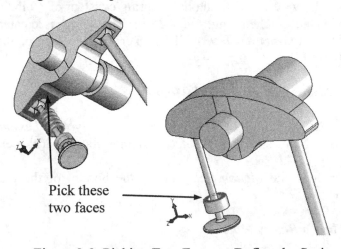

Pick these two faces

Figure 9-9  Picking Two Faces to Define the Spring

Figure 9-8  The *Spring* Dialog Box

In case you picked a wrong entity, simply select the entire text in the respective text field in the dialog box, and press the *Delete* key to delete the text. The text field will be cleaned up and ready for you to pick other entities.

In the *Spring* dialog box, enter the following:

*Stiffness: 10*
*Length: 1.25*
*Coil Diameter: 0.75*
*Number of coils: 8*
*Wire Diameter: 0.1*

Click *the checkmark* ✔ to accept the definition and close the *Spring* dialog box. A spring node (*LinearSpring1*) should appear in the *MotionManager* tree.

### Defining and Running Simulation

Choose *Motion Analysis* for the study. Click the *Motion Study Properties* button ⚙ from the *Motion* toolbar. In the *Motion Study Properties* dialog box, enter *1000* for *Frames per second*, and click the checkmark on top. Zoom in the timeline area until you can see tenth second marks. Drag the end time key to 0.5 second mark in the timeline area to define the simulation duration.

Click *Calculate* button 📊 from the *Motion* toolbar to simulate the motion. A 0.5 second simulation will be carried out. After a few seconds, you should see the camshaft starts rotating, the pushrod is moving up and down, which drives the rocker and then the valve. The camshaft rotates *5* times in the *0.5-second* simulation duration. You may want to adjust the playback speed to 10% to slow down the animation since the total simulation duration is only 0.5 seconds. The motion looks good.

We will graph the position, velocity, and acceleration of the valve next.

### Displaying Simulation Results

Right click the *valve-1* node in the *MotionManager* tree, and choose *Create Motion Plot*. In the *Results* dialog box, choose *Displacement/Velocity/Acceleration*, select *Linear Displacement* and then *Y Component*. Click the checkmark.

A graph like that of Figure 9-10 should appear, where the valve is moving between *−2.49* and *−2.23* in., traveling about *0.26* in., which is about what was expected, as discussed in Section 9.2.

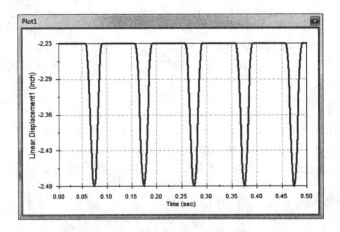

Figure 9-10 Graph of Valve Position

As shown in Figure 9-10, the flat portion on top indicates that the valve stays completely closed, which spans about *0.066* seconds, approximately *240* degrees of the camshaft rotation in a complete cycle. Therefore, the valve will open for about *0.034* seconds per cycle, roughly *120* degrees.

Graph the *Y*-velocity and *Y*-acceleration of the valve by choosing *Displacement/Velocity/Acceleration*, select *Linear Velocity* and then *Y Component* (and *Linear Acceleration*, then *Y Component*). Click the checkmark. The graphs of the velocity and acceleration are shown in Figures 9-11 and 12, respectively. As shown in Figure 9-11, there are two velocity spikes per cycle, representing that the valve is pushed downward (negative velocity) for opening and is being pulled back (positive velocity) for closing, respectively. The valve stays closed with zero velocity.

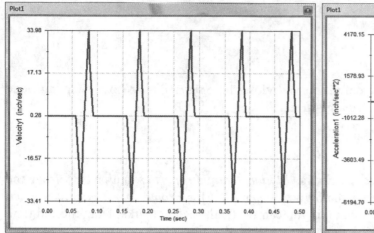

Figure 9-11  Graph of Valve Velocity

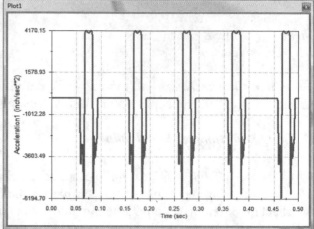

Figure 9-12  Graph of Valve Acceleration

Figure 9-12 reveals high accelerations when the valve is pushed and pulled. Note that such a high acceleration is due to high-speed rotation at the camshaft. This high acceleration could produce large inertial force on the valve, yielding high contact force between the top of the valve and the socket surface in the rocker. We would like to check the reaction force between the top of the valve and the rocker. The graph of the reaction force can be created by expanding the *Mates* node in *MotionManager* tree, right clicking *Concentric3* (between the valve and the rocker), and choosing *Create Motion Plot*. In the *Results* dialog box, choose *Forces*, *Reaction Force*, and then *Y Component*. Click the checkmark. *Motion* shows a warning message, indicating that "this motion study has redundant constraints which can lead to invalid force results. Would you like to replace redundant constraints with bushings…" We will choose *No* for this case since the redundant constraint does not affect this particular result.

The reaction force graph (Figure 9-13) shows that the reaction force between the top of the valve and the socket face of the rocker is about *0.4* lb$_f$, which is insignificant. Note that this small reaction force can be attributed to the small mass of the valve. If you open the valve part and acquire its mass (from pull-down menu, choose *Tools > Evaluate > Mass Properties*), the mass of the valve is *0.03* lb$_m$. Therefore, the inertia for the valve at the peak accelerations is about *0.03* lb$_m$×*4,200* in/sec$^2$ = *126* lb$_m$ in/sec$^2$ = *126/386* lb$_f$ = *0.33* lb$_f$, which is close to peaks found in Figure 9-13.

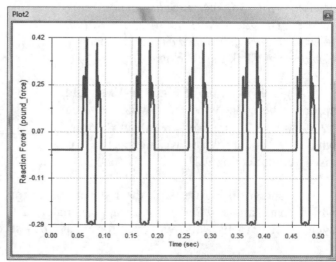

Figure 9-13  Graph of the Reaction Force (*Concentric3*)

If you are not quite sure about why this *386* is factored in for force calculation, please refer to Appendix B for mass and force unit conversions.

Save the model.

**Exercises:**

1.  Redesign the cam by reducing the small arc radius from *0.25* to *0.2* and reducing the center distance of the small arc from *0.52* to *0.40*, as shown in Figure E9-1. Repeat the dynamic analysis and check reaction force between the valve and the rocker. Does this redesigned cam alter the reaction force?

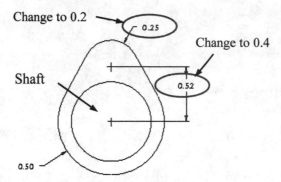

Figure E9-1  The Cam Profile

2.  If we change the *Parallel1* mate between the *Right* plane of the valve and the *Right* plane of the first bushing to a distance mate, will the mechanism move? What other changes must be made in order to create a valid and movable mechanism similar to what was presented in this lesson?

## Lesson 10: Kinematic Analysis of a Racecar Suspension

### 10.1 Overview of the Lesson

In this lesson, we will take a quarter of a racecar suspension and create a motion model for kinematic analysis. The racecar model employed, as shown in Figure 10-1, is a Formula SAE (Society of Automotive Engineers) style racecar designed and built by engineering students at the University of Oklahoma (OU) during 2005-2006. Each year engineering students throughout the world design and build formula-style racecars and participate in the annual Formula SAE competitions (students.sae.org/cds/formulaseries). The result is a great experience for young engineers working on a meaningful engineering project as well as an opportunity to work in a dedicated team environment.

    (a) Manufactured Racecar on Display          (b) Racecar Designed in *Creo*

Figure 10-1  Formula SAE Racecar Designed and Built by OU Engineering Students

The suspension of the entire racecar was modeled for both kinematic and dynamic analyses during 2005-2006. These analysis results were validated using experimental data. The experimental data were acquired by mounting a data acquisition system on the racecar and driving the racecar on a test track following specific driving scenarios that are consistent with those of the simulations. These results were used to aid the suspension design for handling and cornering. Assembling an entire vehicle suspension for motion analysis is non-trivial and is beyond the scope of this book. Therefore, only the right front quarter of the racecar suspension, as shown in Figure 10-2, will be employed in this lesson. The purpose of this lesson is mainly to show you that *Motion* is capable of supporting design of kinematic characteristics of vehicle suspension. Instead of repeating the detailed steps of constructing the motion model in *SOLIDWORKS*, in this lesson, we will start with an assembled motion model. The only component we will add to the motion model is the road profile. The road profile is characterized by the geometric shape of a profile cam, which will be assembled to the tire using a cam-follower assembly mate in *SOLIDWORKS*.

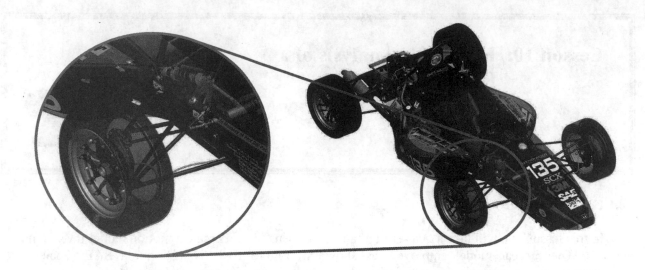

Figure 10-2  The Right Front Quarter of the Racecar Suspension (View A)

## 10.2 The Quarter Suspension

*Physical Model*

The quarter suspension consists of major components that essentially define the kinematic and dynamic characteristics of the racecar. These components include upper and lower control arms, upright, rocker, shock, push rod, tie rod, and wheel and tire, as shown in Figure 10-3. The dangling end of the shock, both control arms, rocker, and tie rod are connected to the chassis frame using numerous joints. The chassis frame is assumed fixed and the tire is pushed and pulled by the profile cam (not shown) mimicking the road profile. Two saved views, *View A* and *View B,* shown in Figure 10-3, are created in the assembled model and will be used for illustrations throughout this lesson.

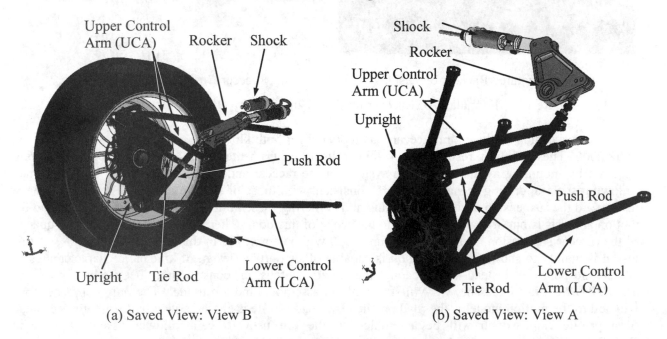

(a) Saved View: View B                    (b) Saved View: View A

Figure 10-3  Major Components of the Quarter Suspension

The tire of the quarter suspension will be in contact with the profile cam representing the road profile. As shown in Figure 10-4, the geometry of the cam consists of two circular arcs of radius 6.65 in. (AB and FG), which are concentric with the cam center. Therefore, when the cam rotates, these two circular arcs do not push or pull the tire, resulting in two flat segments of the road profile, as shown in Figure 10-5. In addition, the circular arc CDE is centered 4 in. above the cam center with a radius of 4 in. Therefore, when the cam rotates, arc CDE pushes the tire up, mimicking a hump of 1.35 in. (that is 1.35 = 8–6.65, peak at point D). A ditch is characterized by an 8 in. arc (HIJ) centered at 3 in. above the cam center. As the cam rotates, arc HIJ creates a ditch of 1.65 in. deep (that is 1.65 = 6.65–(8–3)). The remaining straight lines and arcs provide smooth transitions among flats, humps, and ditches in the road profile.

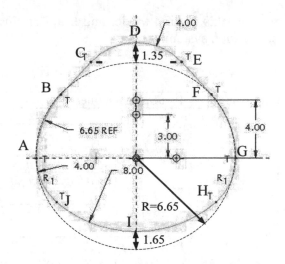

Figure 10-4  Geometry of the Profile Cam

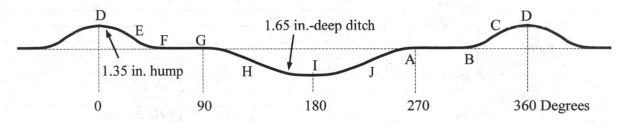

Figure 10-5  Road Profile Generated by the Profile Cam

Based on the geometry of the profile cam, this quarter suspension will go over a 1.35 in. hump and a 1.65 in. ditch in one complete rotation of the profile cam. Note that since the radius of arc AB is 6.65 in., the cam will cause the quarter suspension to travel roughly 41.8 in. (3.48 ft.) in one complete rotation. Since the profile cam will rotate a complete cycle in one second, the suspension travels about 3.48 ft/sec; i.e., 2.37 MPH, a very slow motion.

Note that the in-lb$_f$-sec unit system has been employed for all parts and assemblies. Plain carbon steel is assumed for all parts.

### SOLIDWORKS Parts and Assembly

All *SOLIDWORKS* parts and assemblies are provided for this lesson. In addition, all parts and subassemblies are assembled, except for the profile cam (part name: *profile.sldprt*). The profile cam will be assembled to the tire using a cam-follower connection. A servomotor will be added to drive the cam at a constant angular velocity of 360 degrees/sec, therefore pushing and pulling the tire along the vertical direction, mimicking the situation where the racecar goes over humps and ditches.

Note that the example files you downloaded from the publisher's web site should consist of 23 files: 17 parts, and 6 assemblies, as listed in Table 10-1. The quarter suspension *quarter_suspension.sldasm* is completely assembled except for the road profile (*profile.sldprt*). We will start with this assembly and bring in the profile cam, which will be assembled to the tire using a cam-follower connection. In addition, a completely assembled motion model, *Kinematic.sldasm*, is included for your reference. You may want to open this motion model and bring up this result file to preview the motion animation of the quarter suspension system.

Table 10-1  List of Files in Lesson 10 Folder

| Assembly | Part/Subassemblies | | | Remarks |
|---|---|---|---|---|
| *quarter_suspension.sldasm* | | | | Assembly to start the lesson |
| | *hardpoints.sldprt* | | | Datum features |
| | *fr_rocker_r1.sldasm* | | | Rocker |
| | | *Rocker_Redesign.sldprt* | | |
| | | *Bearing.sldprt (2)* | | |
| | *shock_upper.sldprt* | | | Shocks |
| | *shock_lower.sldprt* | | | |
| | *lca.sldprt* | | | Lower Control Arm |
| | *prod.sldasm* | | | Push Rod |
| | | *prod_Redesign_L.sldprt* | | |
| | | *prod_Redesign_M.sldprt* | | |
| | | *prod_Redesign_S.sldprt* | | |
| | *wheel_r1.sldasm* | | | Wheel |
| | | *wheel.sldprt* | | |
| | | *fr_upright_r1.asm* | | |
| | | | *RH_upright* | |
| | | | *RH_upright_slim* | |
| | | | *RH_steer_arm* | |
| | *uca.sldprt* | | | Upper Control Arm |
| | *trod.sldprt* | | | Tie Rod |
| | *wheel_ref.sldprt* | | | Reference part |
| | *profile.sldprt* | | | To be assembled |
| *Kinematic.sldasm* | | | | Complete Motion Model |

Figure 10-6  Assembly Mates Listed in the Browser

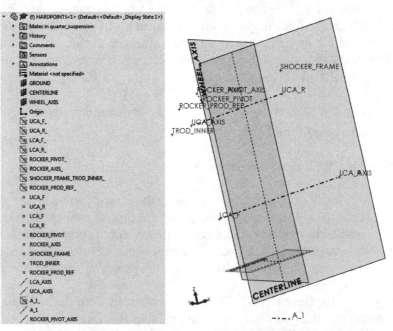

Figure 10-7  The Reference Geometric Entities (or Datum Features) Included in *HARDPOINTS* Employed for Defining Assembly Mates (View B)

There are nineteen assembly mates, including sixteen coincident, one concentric, one angle, and one parallel, as listed in the browser (see Figure 10-6). You may want to expand the *Mates* node in *SOLIDWORKS* browser to review the list of assembly mates. Click any of the mates; you should see the entities selected for the assembly mate highlighted in the graphics area.

The first component, *HARDPOINTS*, shown in the browser is anchored to the assembly. As a result, *HARDPOINTS* is completely fixed to the assembly. As shown in Figure 10-7, there are three planes (*GROUND*, *CENTERLINE*, and *WHEEL_AXIS*), nine 3D axes (*UCA_F_*, *UCA_R_*, *LCA_F_*, *LCA_R_*, *ROCKET_PIVOT_*, *ROCKER_AXIS_*, *SHOCKER_FRAME_TROD_INNER_*, *ROCKER_PROD_REF_*, and *A_1_*), nine points (*UCA_F*, *UCA_R*, *LCA_F*, *LCA_R*, *ROCKET_PIVOT*, *ROCKER_AXIS*, *SHOCKER_FRAME*, *TROD_INNER*, and *ROCKER_PROD_REF*), and four axes (*LCA_AXIS*, *UCA_AXIS*, *A_1*, and *ROCKER_PIVOT_AXIS*) defined in *HARDPOINTS* for support of defining assembly mates.

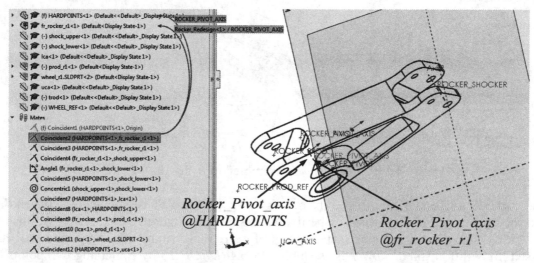

(a) *fr_rocker_r1: Coincident2 (Axes)*

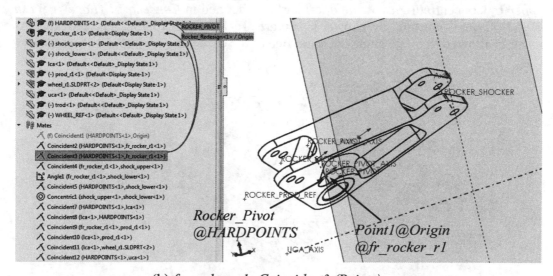

(b) *fr_rocker_r1: Coincident3 (Points)*

Figure 10-8 *fr_rocker_r1* assembled to *HARDPOINTS* (View B)

The second component, *fr_rocker_r1*, is assembled to *HARDPOINTS* by coinciding two axes defined in *Coincident2* (*Rocker_Pivot_axis@fr_rocker_r1* and *Rocker_Pivot_axis@HARDPOINTS*) and coinciding two points defined in *Coincident3* (*Point1@Origin@fr_rocker_r1* and *Rocker_Pivot@HARDPOINTS*), as shown in Figure 10-8. As a result, *fr_rocker_r1* is allowed to rotate along the common axis, which is part of the *HARDPOINTS* and is stationary.

The next component, *shock_upper*, is assembled to *fr_rocker_r1* by coinciding two points, defined in *Coincident4* (*Shocker_rocker@shock_upper* and *Rocker_Shocker@fr_rocker_r1*), as shown in Figure 10-9. As a result, *shock_upper* is allowed to rotate in all three directions at the coincident point.

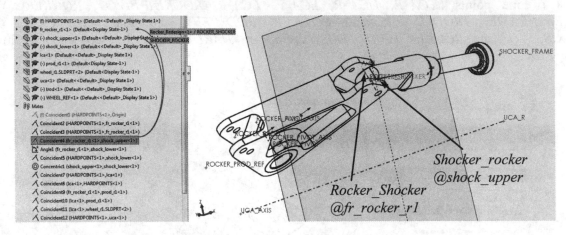

Figure 10-9  *shock_upper* assembled to *fr_rocker_r1* (View B)

Next, *shock_lower* is assembled to the rocker by specifying a 65 degree angle between the two flat faces on the *rocker* and *shock_lower*, respectively, as shown in Figure 10-10a. Note that this angle mate defines initial configuration of the suspension. Before running simulation, this mate will have to be suppressed. In addition, *shock_lower* is assembled to *HARDPOINTS* by coinciding two points defined in *Coincident5* (*Shocker_frame@shock_lower* and *Shocker_frame@HARDPOINTS*), and assembled to *shock_upper* by concentrating two cylindrical surfaces defined in *Concentric1* (*Face<1>@shock_upper* and *Face<2>@shock_lower*), as shown in Figure 10-10b and 10-10c, respectively. As a result, *shock_lower* is allowed to rotate along the centerline of the concentric surfaces (with *Angle1* suppressed).

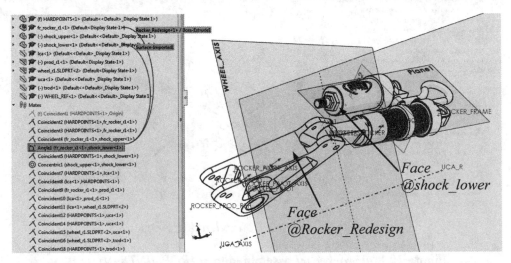

(a) *shock_lower* assembled to *Rocker*: *Angle1 (Faces)*

Figure 10-10  *shock_lower* assembled

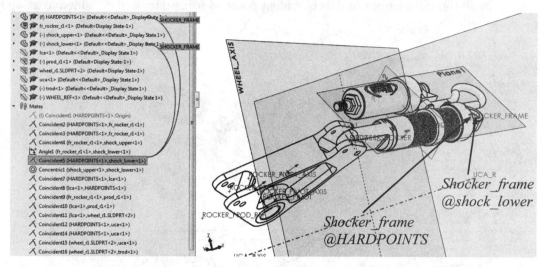

(b) *shock_lower* assembled to *HARDPOINTS*: *Coincident5 (Points)*

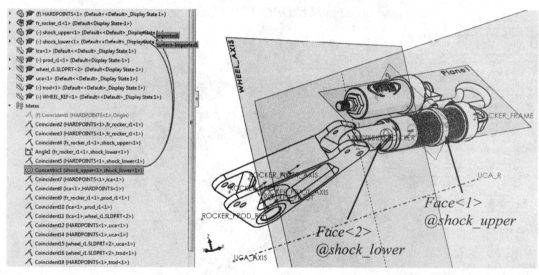

(c) *shock_lower* assembled to *shock_upper*: *Concentric1 (Cylindrical Surfaces)*

Figure 10-10 *shock_lower* assembled (View B) (cont'd)

Next, *lca* (lower control arm) is assembled to *HARDPOINTS* by coinciding two axes defined in *Coincident7* (*Lca_axis@lca* and *Lca_axis@HARDPOINTS*), and coinciding two points defined in *Coincident8* (*Lca_fr@lca* and *Lca_f@HARDPOINTS*), as shown in Figure 10-11a and 10-11b, respectively. As a result, *lca* is allowed to rotate along the common axes.

*prod* (push rod) is assembled to *fr_rocker_r1* by coinciding two points defined in *Coincident9* (*Prod_inner@prod* and *Rocker_prod@fr_rocker_r1*), and assembled to *lca* by coinciding two points defined in *Coincident10* (*Prod_outer@prod* and *Lca_prod@lca*), as shown in Figure 10-12a and 10-12b, respectively. As a result, *prod* is allowed to rotate between these two sets of coincident points.

*wheel_r1* is temporarily assembled to *lca* by coinciding two points defined in *Coincident11* (*Upright_lca@wheel_r1* and *Point1@Origin@lca*), as shown in Figure 10-13. As a result, *wheel_r1* is allowed to rotate in all three directions at the coincident points. More mates will be added to assemble the wheel later.

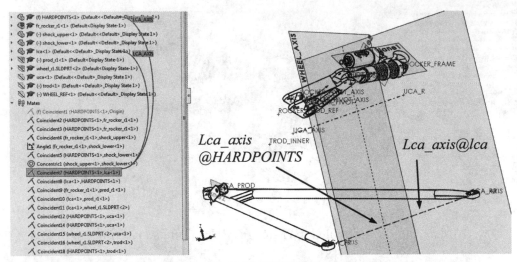

(a) *lca* assembled to *HARDPOINTS*: *Coincident7 (Axes)*

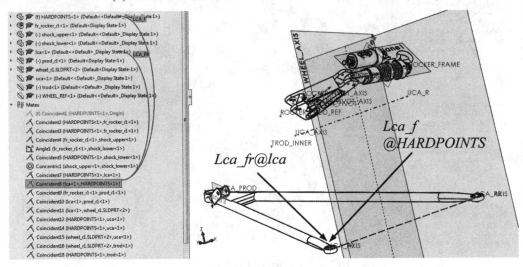

(b) *lca* assembled to *HARDPOINTS*: *Coincident8 (Points)*

Figure 10-11   *lca* assembled (View B)

*uca* (upper control arm) is assembled to *HARDPOINTS* by coinciding two axes defined in *Coincident12* (*Ucr_axis@uca* and *Uca_axis@HARDPOINTS*), two points defined in *Coincident14* (*Ucr_fr@uca* and *Uca_f@HARDPOINTS*), and as shown in Figure 10-14a and 10-14b, respectively. As a result, *uca* is allowed to rotate along the common axes. In addition, *wheel_r1* is assembled to *uca* by coinciding two points defined in *Coincident15* (*Upright_uca@wheel_r1* and *Point1@Origin@uca*), as shown in Figure 10-14c.

Next is the tie rod. *trod* is assembled to *wheel_r1* by coinciding two points defined in *Coincident16* (*Trod_upright@wheel_r1* and *Trod_outer@trod*), as shown in Figure 10-15a. Moreover, the tie rod is assembled to the *HARDPOINTS* by coinciding two points defined in *Coincident16* (*Trod_inner@trod* and *Trod_inner@HARDPOINTS*), as shown in Figure 10-15b. As a result, the *trod* is allowed to rotate in all three directions at the coincident points.

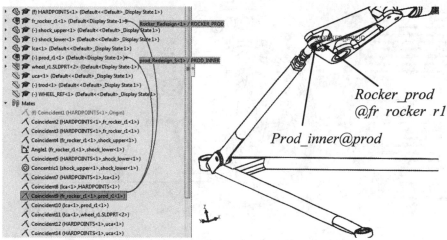

(a) *prod* assembled to *fr_rocker_r1*: *Coincident9 (Points)*

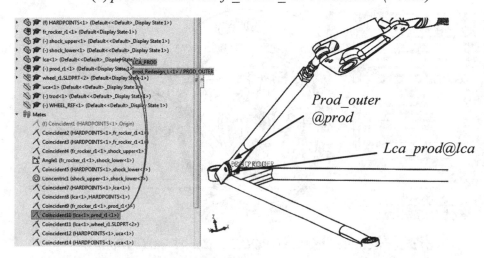

(b) *prod* assembled to *lca*: *Coincident10 (Points)*

Figure 10-12 *prod* assembled (View B)

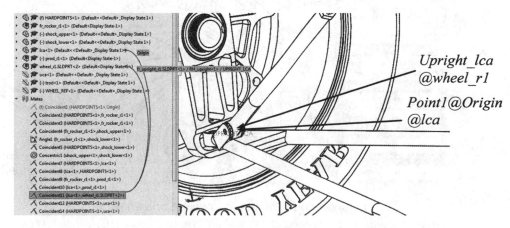

Figure 10-13 *wheel_r1* temporarily assembled to *lca* (View B)

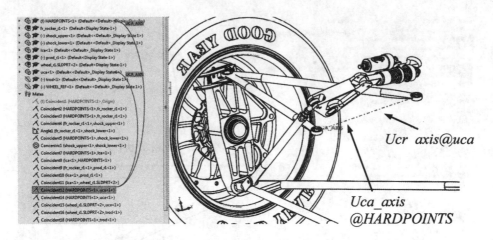

(a) *uca* assembled to *HARDPOINTS: Coincident12 (Axes)*

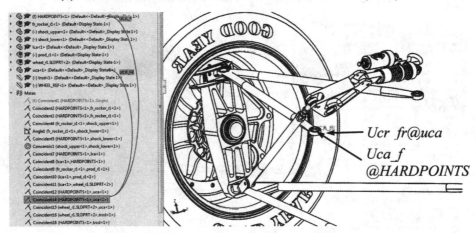

(b) *uca* assembled to *HARDPOINTS: Coincident14 (Points)*

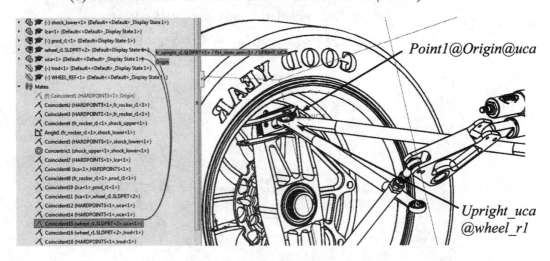

(c) *wheel_r1* assembled to *uca: Coincident15*

Figure 10-14 *uca* assembled (View B)

(a) *trod* assembled to *wheel_r1: Coincident16 (Points)*

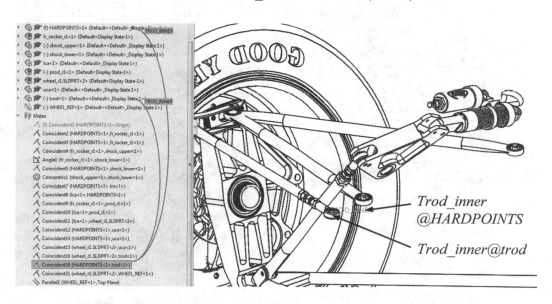

(b) *trod* assembled to *HARDPOINTS: Coincident18 (Points)*

Figure 10-15   *trod* assembled (View B)

Now, we define two more assembly mates for the *wheel_ref*. Note that *wheel_ref* is a part that is created as a patch to aid in defining a cam and follow mate between the wheel and the road profile to be discussed shortly. The *wheel_ref* is assembled to *wheel_r1* by coinciding two points defined in *Coincident21* (*wheel_center@wheel_ref* and *Point1@Origin@wheel_r1*), and assembled to the *Top Plane* by keeping a face and the *Top Plane* in parallel defined in *Parallel2* (*Face<1>@wheel_ref* and *Top Plane*), and as shown in Figure 10-16a and 10-16b, respectively.

After all the nineteen assembly mates are defined, the quarter suspension assembly is completely assembled as shown in Figure 10-17 with the standard three views and a saved view (View B). Note that the assembly is fully constrained. You will have to suppress *Angle1* to allow for motion between components in the suspension assembly. If you suppress *Angle1*, click the wheel and drag it up and down; you will see the movements of individual parts. You may choose from the pull-down menu *Edit > Undo Move Component* to bring the assembly to its original configuration.

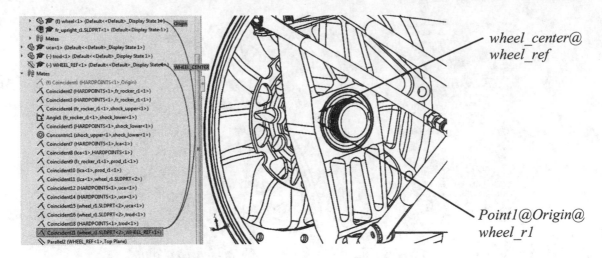

*wheel_center@*
*wheel_ref*

*Point1@Origin@*
*wheel_r1*

(a) *wheel_ref* assembled to *wheel_r1: Coincident21 (Points)*

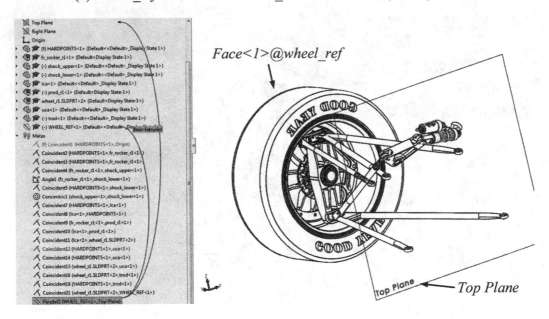

*Face<1>@wheel_ref*

*Top Plane*

(b) *wheel_ref* assembled to *Top Plane: Parallel2 (Planes)*

Figure 10-16 *wheel_ref* assembled (View B)

### Simulation Model

From motion simulation perspective, there are nine bodies defined in this motion model, including the ground body. All the key datum features, including datum coordinate systems, datum axes, and datum points, required for assembly are collected in *HARDPOINTS*, as shown in Figure 10-7. The *HARDPOINTS* was assembled to the assembly *quarter_suspension*. Therefore, *HARDPOINTS* belongs to the ground body. Note that the *X*-axis of the reference triad points to the backward direction of the suspension, as shown in Figure 10-7, and the wheel is about 2.4 in. above the ground in the *Z*-direction (that is the distance between the lowest point in the tire and the reference plane *GROUND*, as shown in Figure 10-18). Note that the distance is measured by choosing *Tools > Evaluate > Measure* from the pull-down menu.

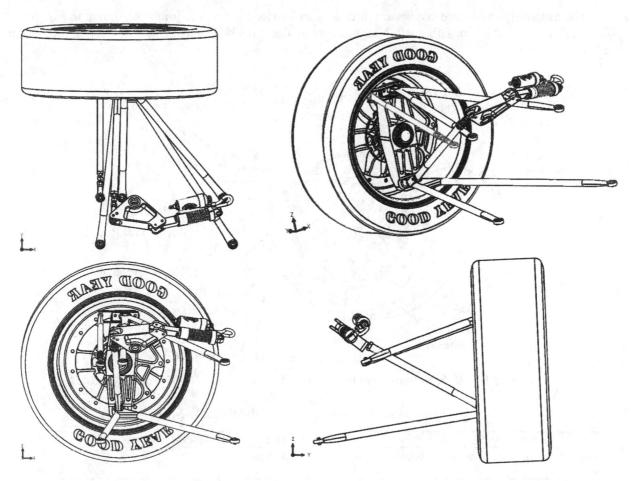

Figure 10-17  The Quarter Suspension Assembly in Standard Three Views and a Prescribed View B
(upper right)

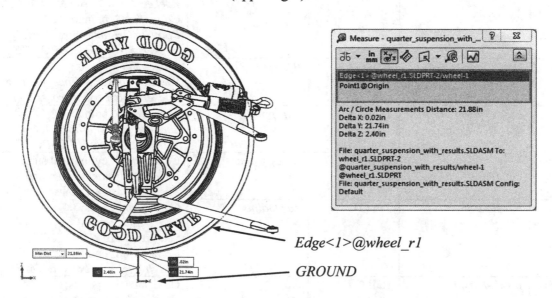

Figure 10-18  A 2.4 in. Distance Between the Lowest Point of Wheel and the *GROUND (*Origin of the
*quarter_suspension* Assembly)*

The assembly mates are converted internally to fourteen kinematic joints, as shown in Figure 10-19. These joints are listed in Table 10-2 together with the nine bodies and their corresponding assembly mates.

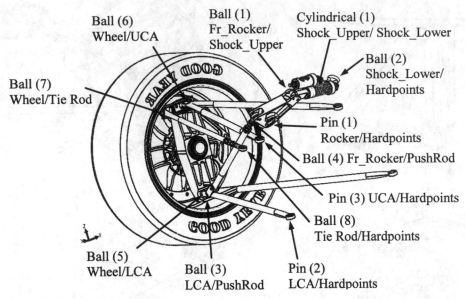

**Figure 10-19 Kinematic Joints Converted from the Assembly Mates (View B)**

**Table 10-2 Kinematic Joints and Assembly Mates**

| Body | Part or Assembly | Joints | Assembled to | Assembly Mates |
|------|------------------|--------|--------------|----------------|
| Ground Body | *hardpoints.sldprt* | Rigid | *quarter_suspension* | |
| Body1 | *ft_rocker_r1.sldasm* | Pin (1) | *hardpoints* | *Coincident2*, axes: *Rocker_Pivot_axis@fr_rocker_r1* and *Rocker_Pivot_axis@HARDPOINTS* *Coincident3*, points: *Point1@Origin@fr_rocker_r1* and *Rocker_Pivot@HARDPOINTS* |
| Body2 | *shock_upper.sldprt* | Ball (1) | *ft_rocker_r1* | *Coincident4*, points: *Shocker_rocker@shock_upper* and *Rocker_Shocker@fr_rocker_r1* |
| Body3 | *shock_lower.sldprt* | Ball (2) | *hardpoints* | *Coincident5*, points: *Shocker_frame@shock_lower* and *Shocker_frame@HARDPOINTS* |
| | | Cylinder (1) | *shock_upper* | *Concentric1*, cylindrical surfaces: *Face<1>@shock_upper* and *Face<2>@shock_lower* |
| Body4 | *lca.sldasm* | Pin (2) | *hardpoints* | *Coincident7*, axes: *Lca_axis@lca* and *Lca_axis@HARDPOINTS* *Coincident8*, points: *Lca_fr@lca* and *Lca_f@HARDPOINTS* |
| Body5 | *prod.sldasm* | Ball (3) | *lca* | *Coincident10*, points: *Prod_outer@prod* and *Lca_prod@lca* |
| | | Ball (4) | *fr_rocker_r1* | *Coincident9*, points: *Prod_inner@prod* and *Rocker_prod@fr_rocker_r1* |

Table 10-2  Kinematic Joints and Assembly Mates (cont'd)

| Body | Part or Assembly | Joints | Assembled to | Assembly Mates |
|------|------------------|--------|--------------|----------------|
| Body6 | *wheel_r1.sldasm* | Ball (5) | *lca* | *Coincident11*, points:<br>*Upright_lca@wheel_r1* and<br>*Point1@Origin@lca* |
| | | Ball (6) | *uca* | *Coincident15*, points:<br>*Upright_uca@wheel_r1* and<br>*Point1@Origin@uca* |
| | | Ball (7) | *trod* | *Coincident16*, points:<br>*Trod_upright@wheel_r1* and<br>*Trod_outer@trod* |
| Body7 | *uca.sldasm* | Pin (3) | *hardpoints* | *Coincident12*, axes:<br>*Ucr_axis@uca* and<br>*Uca_axis@HARDPOINTS*<br>*Coincident14*, points:<br>*Ucr_fr@uca* and *Uca_f@HARDPOINTS* |
| Body8 | *trod.sldasm* | Ball (8) | *hardpoints* | *Coincident18*, points:<br>*Trod_inner@trod* and<br>*Trod_inner@HARDPOINTS* |

## 10.3 Using *SOLIDWORKS Motion*

### *Assembling the Profile Cam*

Start *SOLIDWORKS* and open *quarter_suspension.sldasm*. You should see *quarter_suspension.sldasm* appear in the graphics area with default view *View B* (see Figure 10-3a). Note that the *quarter_suspension.sldasm* consists of nine components, listed in the browser; i.e., *HARDPOINTS*, *fr_rocker_r1*, *shock_upper*, *shock_lower*, *lca*, *prod*, *wheel_r1.sldasm*, *uca*, and *trod.sldasm*. You may choose the *Configuration* tab 🔳 above the browser, right click *ExpView1* and choose *Explode* to see the explode view of the assembly, as shown in Figure 10-20. Right click *ExpView1* and choose *Collapse* to restore the unexplode view of the assembly. Check the units system and make sure that *IPS* units system is chosen for this example.

Figure 10-20  Displaying the Explode View Under the *Configuration* Tab

To add the cam to the assembly, click *Insert > Component > Existing Part/Assembly* from the pull-down menu and click *Browse* in *Insert Component* dialog box to insert the part *profile*. Place the profile cam at an arbitrary location in the graphics area. Click *Insert > Mate* from the pull-down menu to bring up the *Mate* dialog box (Figure 10-21). Pick reference axis *A_1* (*profile*) and reference axis

*A_1(HARDPOINTS)* from the feature tree (Figure 10-22). Make sure the mate type *Coincide* is selected, and then click ✓ on top of the dialog box to accept the mate.

Pick the *Front Plane* in *profile* and the *CENTERLINE* plane in *HARDPOINTS*, then click the *Distance* button in the *Mate* dialog box. Enter *24.5* and press the *Enter* key (See Figure 10-23). You may need to check or cancel the *Flip dimension* option to obtain the correct mate as shown in Figure 10-23. Click the checkmark ✓ to accept the definition.

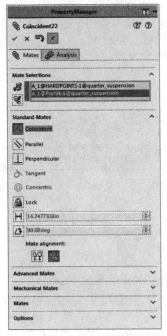

Figure 10-21  The *Mate* Dialog Box

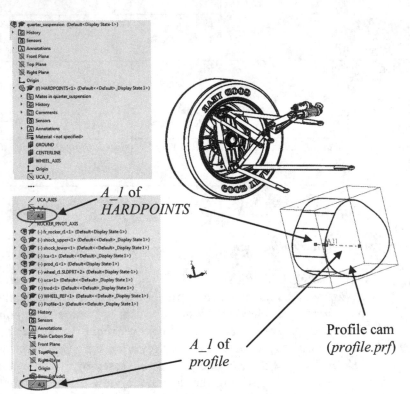

Figure 10-22  Picking the Axes for Assembly

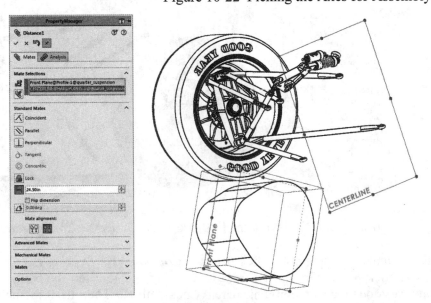

Figure 10-23  Picking Planes for Defining a Distance Mate

In order to fix the rotation degree of freedom of the cam, first pick the *Top Plane* of *profile* and the *GROUND* plane in *HARDPOINTS*, as shown in Figure 10-24. The *Parallel* mate is automatically selected, and the orientation of the cam should be similar to that shown in Figure 10-24, which means we want the hump of the profile cam to point downward at the beginning of the simulation. Click the checkmark ✅ twice to close the *Mate* dialog box.

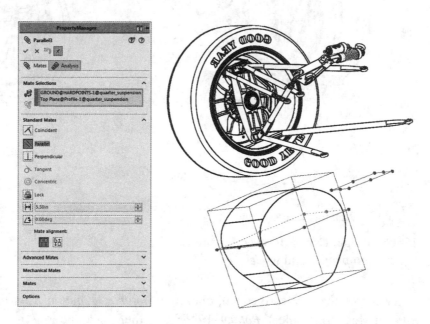

Figure 10-24　Picking Planes for Defining a Parallel Mate (*Parallel3*)

The next step is to create a connection between the tire and the profile cam. However, in *SOLIDWORKS*, it is not allowed to define a cam mate in which the axis of the cylindrical surface of the follower is not parallel with the axis of the cam. Therefore, we need to bring in an auxiliary component, named *wheel_ref*, which acts as the follower. The cam mate can be defined between the wheel_ref and the profile cam, meanwhile the wheel_ref is connected to the wheel at a given point so that the motion due to cam rotation can be transmitted to the wheel through the wheel_ref.

Before defining a cam mate, we suppress two mates to provide adequate freedom and avoid conflict in the assembly. These two mates are *Angle1 (fr_rocker_r1<1>,shock_lower<1>)* and *Parallel3 (HARDPOINTS<1>,profile<1>)*. Expand the *Mates* 🔗 in the *MotionManager* tree. Right click *Angle1* and choose *Suppress*, and then Right click *Parallel3* and choose *Suppress*.

From the pull-down menu, choose *Insert > Mate*. In the *Mate* dialog box, expand the *Mechanical Mates* label and click *Cam* 📷. In the graphics area, select a surface in *profile* for the *Cam Path* box, and then move the pointer closer to the outer surface of the tire until a label *Boss-Extrude of WHEEL_REF<1>* appears (see Figure 10-25). Click the surface for *Cam Follower* box; *Face<3>@WHEEL_REF-1* appears in the *Cam Follower* box. In the graphics area, the profile and wheel_ref should come into contact like that of Figure 10-26. Click the checkmark ✅ twice to close the dialog box (click *OK* if a warning message appears).

### *Creating a Motion Model*

Now the motion model has been completely assembled. The next step is to set up a kinematic analysis using *SOLIDWORKS Motion*. Click the *Motion Study 1* tab at the bottom of the graphics area to start a new study. Right click the tab and rename the motion study as *Kinematic*.

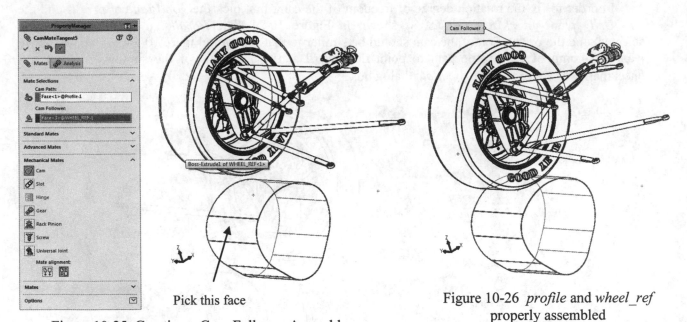

Pick this face

Figure 10-25  Creating a Cam-Follower Assembly
Mate between *profile* and *wheel_ref*

Figure 10-26  *profile* and *wheel_ref*
properly assembled

To add a rotary motor that drives the cam, click the *Motor* ![icon] button from the *Motion* toolbar to bring up the *Motor* dialog box. Select *Rotary Motor* ![icon], then pick the *A_1* axis of the profile cam (*profile<1>*) for *Component/Direction*. A red circular arrow showing the rotational direction will appear, as shown in Figure 10-27. Click ![icon] to switch the direction if necessary. Choose *Constant speed* and input *60 RPM* as the rotational speed for the cam. Click the ![icon] button on top of the dialog box to accept the motor definition. *RotaryMotor1* can now be found in the *MotionManager* tree.

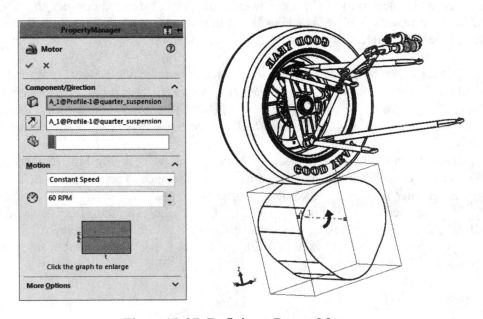

Figure 10-27  Defining a Rotary Motor

### Defining and Running Simulation

Choose *Motion Analysis* for the study. Click ⚙ from the *Motion* toolbar to edit the properties of the study. Enter *500* for *Frames per second* then click ✔ button to accept. Define the duration by dragging the end time key to the *2*-second mark in the timeline area, so that the profile cam will make two complete turns during the simulation. Click the *Calculate* 🖩 button from the *Motion* toolbar to start the simulation. Note that you may need to re-suppress the mates, *Angle1* and/or *Parallel3*, to allow adequate freedom for the system to move. You should see the cam rotating and the wheel being pushed and pulled by the cam. The cam will stop after two complete turns. Click the *Play from Start* ▐▶ button to replay the animation. Also, you can click 🎞 from the *Motion* toolbar to save the animation as an AVI video.

### Displaying Simulation Results

Next, we will create graphs of the simulation results to monitor the three measures: vertical wheel travel, shock travel, and camber angle.

Click the *Results and Plots* 🖳 button from the *Motion* toolbar. In the *Results* dialog box, choose *Displacement/Velocity/Acceleration*, select *Linear Displacement*, and then *Z Component*. In the browser, pick reference point *Measure* in *wheel<1>* (expanding *wheel_r1.SLDPRT* in the browser) and the origin of *HARDPOINTS<1>*. Click the checkmark ✔ to accept the definition, and you should see a graph similar to Figure 10-28. If the scales of the *Y* axis are integers, double click on the *Y* axis scales and the *Format Axis* dialog box should appear. Switch to the *Number* tab, change the *Decimal Places* to *1* and click *OK*.

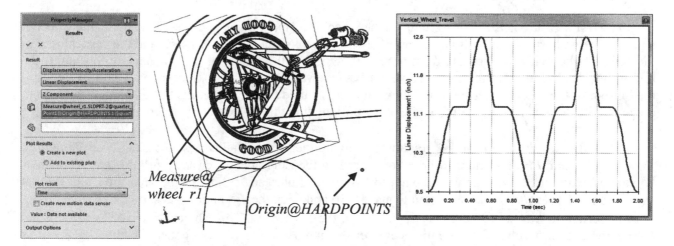

Figure 10-28  Vertical Position of Wheel Center

This graph shows the vertical position of the wheel center as the profile cam rotates. The center of the wheel travels vertically between about 9.5 in. and 12.6 in. The flat portion (11.2 in.) resembles the flat road profile. The distance between the peak and the flat portion is about 1.4 in. due to the 1.35 in. hump. Similarly, the distance between the flat portion and the crest is about 1.7 in. due to the 1.65 in. ditch. Close *Plot1*. In the *MotionManager* tree, you can find the graph by expanding the *Results* folder. Click the name of the graph and press the *F2* key to rename the graph as *Vertical_Wheel_Travel*.

The graph of the other two measures, shock travel and camber angle, can be defined following similar steps. To plot the shock travel, Click 🖳, select *Displacement/Velocity/Acceleration* in the *Results* dialog box, choose *Linear Displacement*, and then *Magnitude*. Pick reference point

*SHOCKER_FRAME* in *HARDPOINTS<1>* and reference point *ROCKER_SHOCKER* in *rocker_redesign<1>* (within the assembly *fr_rocker<1>*) from the browser to define the shock travel. Click ✔ to create the graph, like that of Figure 10-29, which shows that the shock travels between about 5.9 in. and about 8.4 in. The overall travel distance is about 2.5 in., which is probably too large for such a small hump or ditch. In fact, in the simulation, it appears that the shock is compressed too much, in which the piston penetrates into its reserve tube. In reality, this will not happen. However, the simulation raises a flag indicating that there could be severe contact within the shock, leading to potential part failure. Rename the graph as *Shock_Travel* in the *MotionManager* tree.

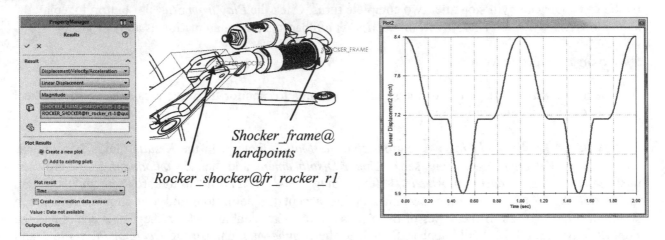

Shocker_frame@
hardpoints

Rocker_shocker@fr_rocker_r1

Figure 10-29  Distance of Shock Travel

The last measure is the camber angle of the wheel. A camber angle is the angle made by the wheel of a ground vehicle; specifically, it is the angle between the vertical axis of the wheel and the vertical axis of the vehicle when viewed from the front or rear. It is used in the design of steering and suspension. If the top of the wheel is further out than the bottom (that is, away from the axle), it is called positive camber; if the bottom of the wheel is further out than the top, it is called negative camber. In this model, the camber angle will be defined as the rotation angle of the upright along the *X*-axis of the reference triad.

Click the *Results and Plots* button to bring up the *Results* dialog box. Choose *Other Quantities*, *Euler Angles*, and *Phi*. For more information regarding the Euler angles, please refer to resources such as Dynamics textbooks or Wikipedia (en.wikipedia.org/wiki/Euler_angles). In the graphics area, pick the flat surface on *RH_upright<1>* as shown in Figure 10-30 and click the ✔ button. As shown in Figure 10-30, the camber angle was set to be about *–1* degree on the flat terrain. The camber angle varies to *–2.5* and *0.4* degrees, respectively, when the tire goes over the hump and the ditch. In general, camber angle alters the handling qualities of a particular suspension design. In general, negative camber improves grip when cornering. This is because it places the tire at a more optimal angle to the road, transmitting the forces through the vertical plane of the tire, rather than through a shear force across it. However, excessive negative camber change in hump can cause early lockup under breaking or wheel spin under acceleration. There is only limited information that can be obtained by conducting kinematic analysis of the quarter suspension. Ultimately, a full-vehicle dynamic simulation must be carried out to fully understand the suspension design and hopefully develop a strategy for design improvement.

### Interference Detection

Before completing the lesson, we want to check if there is any interference between parts when the suspension is in motion. However, there are too many parts in this assembly. Interference checking we learned in *Lesson 6* requires a lot more time for *Motion* to perform the calculation. Instead, we use the interference detection capability provided in the standard assembly mode of *SOLIDWORKS* to just identify if there is any major concern in part penetration at the current configuration.

From the pull-down menu, choose *Tools > Evaluate > Interference Detection*. In the *Interference Detection* dialog box (Figure 10-31), click the *Calculate* button. Interferences detected are listed in the *Results* area. Expand interferences detected to learn more, for example, *Interference3*, and click parts with interference in between to see them in the graphics area (Figure 10-31). Note that some of the interferences detected may be simply modeling errors. However, some may reveal significant design defects that need to be taken care of before entering the next design phase or manufacturing.

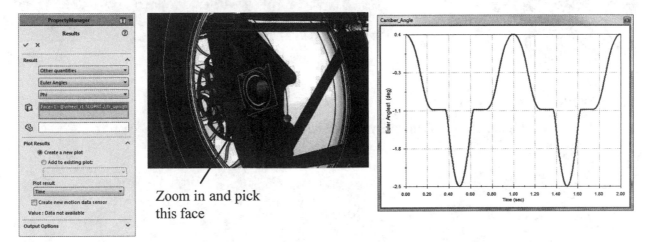

Zoom in and pick this face

Figure 10-30  Camber Angle

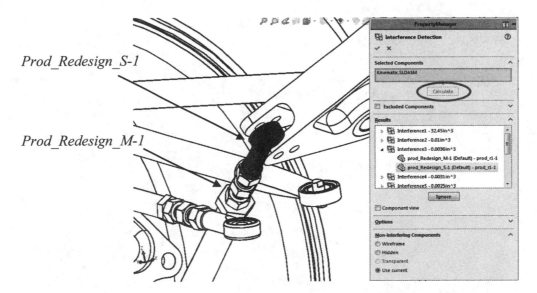

*Prod_Redesign_S-1*

*Prod_Redesign_M-1*

Figure 10-31  Interference Detection

### Full-Vehicle Dynamic Simulation

A final note: a full-vehicle dynamic simulation model was created in *ADAMS/Car*, using the model templates provided, as shown in Figure 10-32a. With *ADAMS/Car*, users can simply enter vehicle model data into the templates, and *ADAMS/Car* will automatically construct subsystem models, such as engine, shock absorbers, tires, as well as the full vehicle assemblies. Once these templates are created, they can be made available to novice users, enabling them to perform standardized vehicle maneuvers. The vehicle model was then simulated for various test scenarios, including skid pad racing, which is a constant radius cornering simulation as shown in Figure 10-32b.

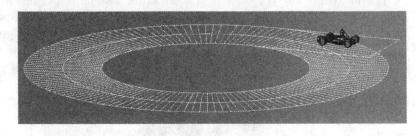

(a) 15 DOFs *ADAMS/Car* Model                                      (b) Skid Pad Racing

Figure 10-32  Vehicle Dynamic Simulation of Formula SAE Racecar

## APPENDIX A: DEFINING JOINTS

**Degrees of Freedom**

Understanding degrees of freedom is critical in creating successful motion models. The free degrees of freedom of a mechanism represent the number of independent parameters required to specify the position, velocity, and acceleration of each rigid body in the system for any given time. A completely unconstrained body in space has six degrees of freedom: three translational and three rotational. If you add a joint; e.g., a revolute joint to the body, you restrict its movement to rotation about an axis, and the free degrees of freedom of the body are reduced from six to one.

For a given motion model, you can determine its number of degrees of freedom using the Gruebler's count. The mechanism's Gruebler count is calculated using the mechanism's total number of bodies, and the numbers and types of joints. As mentioned above, each movable body introduces six degrees of freedom. Joints added to the mechanism constrain the system, or remove DOFs. Motion drivers or forces applied to the system remove additional DOFs.

*SOLIDWORKS Motion* uses the following equation to calculate the Gruebler's count:

$$D = 6M - N - O \tag{A.1}$$

where $D$ is the Gruebler count representing the total free degrees of freedom of the mechanism, $M$ is the number of bodies excluding the ground body, $N$ is the number of DOFs restricted by all joints, and $O$ is the number of the motion inputs (such as a linear or rotary motor) added to the system.

For kinematic analysis, the Gruebler's count must be equal to or less than *0*. The *ADAMS/Solver* recognizes and deactivates redundant constraints during motion simulation. For a kinematic analysis, if you create a model with a Gruebler's count greater than *0* and try to simulate it, the simulation will not run and an error message will appear.

If the Gruebler's count is less than zero, the solver will automatically remove redundancies. For example, you may apply this formula to a door model that is supported by two hinges modeled as revolute joints. Since a revolute joint removes five DOFs, the Gruebler's count becomes:

$$D = (6 \times 1) - (2 \times 5) = -4.$$

The calculated degrees of freedom result is *–4*, which include five redundant DOFs.

**Redundancy**

Redundancies are excessive (or duplicate) DOFs. When a joint constrains the model in exactly the same way as another joint (like the door hinge example), the model contains excessive DOFs, also known as redundancies. A joint becomes excessive when it does not introduce any further restriction on a body's motion.

It is important that you understand the redundancies in the motion model and be mindful in reviewing dynamic simulation results at these redundant DOFs. For example, if you model a door using two revolute joints for the hinges, the second revolute joint does not contribute to constraining the door's motion. *Motion* solver detects the redundancies and ignores one of the resolute joints in its analysis. The outcome on the redundant DOFs may be incorrectly reported in reaction results, yet the motion is correct.

For a kinematic simulation where you are interested in displacement, velocity, and acceleration, redundancies in your model do not alter the performance of the mechanism.

You can eliminate or reduce the redundancies in your model by carefully choosing joints. These joints must be able to restrict the same DOFs but not duplicate each other, introducing redundancies. After you decide which joints you want to use, you can use the Gruebler's count to calculate the DOFs and check redundancies.

Also, after carrying out a motion analysis, *Motion* will show number of redundancies in the *MotionManager* tree, as shown in Figure A-1. You may ask *Motion* to show you motion analysis message by selecting *Show all Motion Analysis messages* in the *Motion Study Properties* dialog box, as shown in Figure A-2. In the message displayed, *Motion* shows you which redundant DOFs are removed in analysis (please see Figure A-3 for an example).

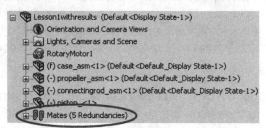

Figure A-1  The *MotionManager* Tree

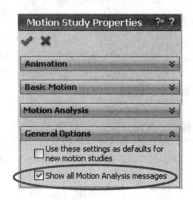

Figure A-2  The *Motion Study Properties* Dialog

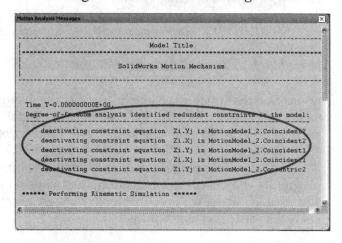

Figure A-3  The Motion Analysis Messages

**Joint Types**

Before you select a joint to add to your model, you should know what movement you want to restrain for the body and what movement you want to allow. Table A-1 describes the commonly employed joint types in dynamic simulations and the degrees of freedom they remove.

The following provides more details about the joints listed in the table.

*Revolute Joint*

A revolute joint, as depicted in Figure A-4, allows the rotation of one rigid body with respect to another rigid body about a common axis. The origin of the revolute joint can be located anywhere along the axis about which the bodies can rotate with respect to each other. Orientation of the revolute joint defines the direction of the axis about which the bodies can rotate with respect to each other. The rotational axis of the revolute joint is parallel to the orientation vector and passes through the origin.

Table A-1  Common Joints Employed for Motion Simulations

| Joint Type | DOF Removed | | | Remarks |
|---|---|---|---|---|
| | Translation | Rotation | Total | |
| Revolute | 3 | 2 | 5 | Rotates about an axis |
| Translational | 2 | 3 | 5 | Translates along an axis |
| Cylindrical | 2 | 2 | 4 | Translates along and rotates about an axis |
| Spherical | 3 | 0 | 3 | Rotates in any direction |
| Universal | 3 | 1 | 4 | Rotates about two axes |
| Screw | 0.5 | 0.5 | 1 | Coupled rotation and translation along one axis |
| Planar | 2 | 1 | 3 | Bodies connected by a planar joint move in a plane with respect to each other. Rotation is about an axis perpendicular to the plane. |
| Fixed | 3 | 3 | 6 | Glues two parts together. Parts constrained by a rigid connection constitute a single body. |

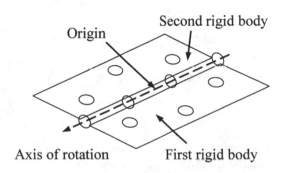

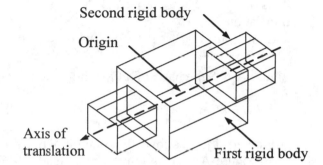

Figure A-4  Schematic of a Revolute Joint        Figure A-5  Schematic of a Translational Joint

*Translational Joint*

A translational joint allows one rigid body to translate along a vector with respect to a second rigid body, as illustrated in Figure A-5. The rigid bodies may only translate, not rotate, with respect to each other.

The location of the origin of a translational joint with respect to its rigid bodies does not affect the motion of the joint but does affect the reaction loads on the joint.

The orientation of the translational joint determines the direction of the axis along which the bodies can slide with respect to each other (axis of translation). The direction of the motion of the translational joint is parallel to the orientation vector and passes through the origin.

*Cylindrical Joint*

A cylindrical joint allows both relative rotation and relative translation of one body with respect to another body, as shown in Figure A-6. The origin of the cylindrical joint can be located anywhere along the axis about which the bodies rotate or slide with respect to each other.

Orientation of the cylindrical joint defines the direction of the axis about which the bodies rotate or slide along with respect to each other. The rotational/translational axis of the cylindrical joint is parallel to the orientation vector and passes through the origin.

*Spherical Joint*

A spherical joint allows free rotation about a common point of one body with respect to another body, as depicted in Figure A-7. The origin location of the spherical joint determines the point about which the bodies pivot freely with respect to each other.

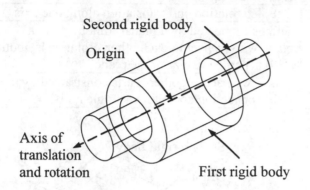

Figure A-6  Schematic of a Cylindrical Joint

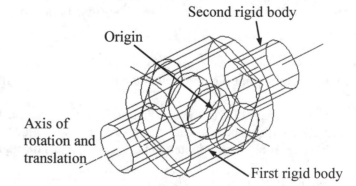

Figure A-7  Schematic of a Spherical Joint

*Universal Joint*

A universal joint allows the rotation of one body to be transferred to the rotation of another body, as shown in Figure A-8. This joint is particularly useful to transfer rotational motion around corners, or to transfer rotational motion between two connected shafts that are permitted to bend at the connection point (such as the drive shaft on an automobile).

The origin location of the universal joint represents the connection point of the two bodies. The two shaft axes identify the center lines of the two bodies connected by the universal joint.

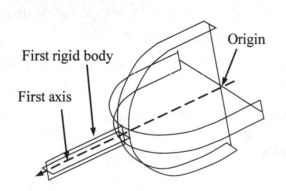

Figure A-8  Schematic of a Universal Joint

Figure A-9  Schematic of a Screw Joint

*Screw Joint*

A screw joint removes one degree of freedom. It constrains one body to rotate as it translates with respect to another body, as shown in Figure A-9.

When defining a screw joint, you can define the pitch. The pitch is the amount of translational displacement of the two bodies for each full rotation of the first body. The displacement of the first body relative to the second body is a function of the body's rotation about the axis of rotation. For every full rotation, the displacement of the first body along the translation axis with respect to the second body is equal to the value of the pitch.

Very often, the screw joint is used with a cylindrical joint. The cylindrical joint removes two translational and two rotational degrees of freedom. The screw joint removes one more degree of freedom by constraining the translational motion to be proportional to the rotational motion.

### *Planar Joint*

A planar joint allows a plane on one body to slide and rotate in the plane of another body, as shown in Figure A-10.

The orientation vector of the planar joint is perpendicular to the joint's plane of motion. The rotational axis of the planar joint, which is normal to the joint's plane of motion, is parallel to the orientation vector.

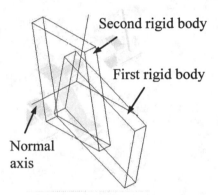

Figure A-10  Schematic of a Planar Joint          Figure A-11  Schematic of a Fixed Joint

### *Fixed Joint*

A fixed joint locks two bodies together so they cannot move with respect to each other. The schematic of a fixed joint is shown in Figure A-11.

## APPENDIX B: THE UNITS SYSTEM

The mass unit $lb_m$ is not quite common to many engineers. The basic physical quantities involved in determining a units system are length, time, mass, and force. These four basic quantities are related through Newton's second law,

$$F = ma \tag{B.1}$$

where $F$, $m$, and $a$ are force, mass, and acceleration (length per second square), respectively.

When the unit $lb_m$ for mass is employed, the force unit will be determined by length (in.), mass ($lb_m$), and second (sec) through Eq. B.1; i.e.,

$$1 \ lb_m \ in/sec^2 \ (force) = 1 \ lb_m \ (mass) \times 1 \ in/sec^2 \ (acceleration) \tag{B.2}$$

where the force unit, $lb_m$ in/sec$^2$, is a derived unit.

From Eq. B.2, a $1$ $lb_m$ in/sec$^2$ force will generate a $1$ in/sec$^2$ acceleration when applied to a $1$ $lb_m$ mass block, as shown in Figure B-1a. The same block will weigh $1$ $lb_f$ on earth (see Figure B-1b), where the gravitational acceleration is assumed $386$ in/sec$^2$; i.e.

$$1 \ lb_f \ (force) = 1 \ lb_m \ (mass) \times 386 \ in/sec^2 \ (acceleration) \tag{B.3}$$

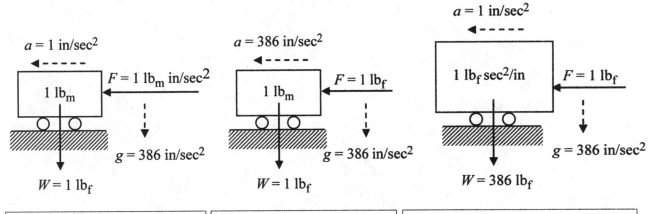

(a) A 1 $lb_m$ in/sec$^2$ Force Applied to a 1 $lb_m$ Mass Block

(b) A 1 $lb_f$ Force Applied to a 1 $lb_m$ Mass Block

(c) A 1 $lb_f$ Force Applied to a 1 $lb_f$ sec$^2$/in Mass Block

Figure B-1  Forces Applied on Blocks of Different Masses

Therefore, from Eqs. B.2 and B.3, we have $1$ $lb_f = 386$ $lb_m$ in/sec$^2$, and the force in $lb_m$ in/sec$^2$ unit is 386 times smaller than that of $lb_f$ that we are more used to. When you apply a $1$ $lb_f$ force to the same mass block, it will accelerate $386$ in/sec$^2$, as shown in Figure B-1b.

On the other hand, we have the mass unit, $1$ $lb_m = 1/386$ $lb_f$ sec$^2$/in. It means that a $1$ $lb_m$ mass block is $386$ times smaller than that of a $1$ $lb_f$ sec$^2$/in block. Therefore, a $1$ $lb_f$ sec$^2$/in block will weigh $386$ $lb_f$ on earth. When applying a $1$ $lb_f$ force to the mass block, it will accelerate at a $1$ in/sec$^2$ rate, as illustrated in Figure B-1c.

The mass unit slug that we are more familiar with is defined as

$$1\ lb_f\ (force) = 1\ slug\ (mass) \times 1\ ft/sec^2\ (acceleration) = 386\ lb_m\ in/sec^2 = 32.2\ lb_m\ ft/sec^2 \qquad \text{(B.4)}$$

Therefore, we have $1$ slug $= 32.2$ $lb_m$.

**APPENDIX C: IMPORTING *Pro/ENGINEER* PARTS AND ASSEMBLIES**

From time to time when you use *Motion* for simulations, you may encounter the need for importing solid models from other CAD software, such as *Pro/ENGINEER*. *SOLIDWORKS* provides an excellent capability that support importing solid models from a broad range of software and formats, including Parasolid, ACIS, IGES (Initial Graphics Exchange Standards), STEP (STandard for Exchange of Product data), *SolidEdge*, *Pro/ENGINEER*, etc. For a complete list of supported software and formats in *SOLIDWORKS*, please refer to Figure C-1. You may access this list by choosing *File > Open* from the pull-down menu, and pull down the *Files of type* in the *File Open* dialog box.

In this appendix, we will focus on importing *Pro/ENGINEER* parts and assemblies. Hopefully, methods and principles you learn from this appendix will be applicable to importing solid models from other software and formats.

*SOLIDWORKS* provides capabilities for importing both part and assembly. Users can choose two options in importing solid model. They are Option 1: importing solid features and Option 2: importing just geometry. Importing solid features may bring you a parametric solid model that you will be able to modify just like a *SOLIDWORKS* part. On the other hand, if you choose to import geometry only, you will end up with an imported feature that you cannot change since all solid features are lumped into a single imported geometry without any solid features or dimensions.

Figure C-1 *Open* Dialog Box

Importing geometry is relatively straightforward. In general, *SOLIDWORKS* does a good job bringing in a *Pro/ENGINEER* part as a single imported geometry. In fact, several other translators, such as IGES and STEP, support such geometric translations well. IGES and STEP are especially useful when there is no direct translation from one CAD to another.

Importing solid models with solid features is a lot more challenging because solid features embedded in the part geometry, such as holes, chamfers, etc., must be identified first. In addition, sketches that were employed for generating the solid features must be recovered and the feature types, for instance revolve, extrude, sweep, etc., must be identified. With a virtually infinite number of possibilities in creating solid features, it is almost certain that you will encounter problems while importing solid models with feature conversion. Therefore, if you do not anticipate making design changes in *SOLIDWORKS*, it is highly recommended that you import parts as a single geometric feature.

We will discuss the approaches of importing parts and then importing assemblies. In each case, we will try both options; i.e., importing solid features vs. importing geometry. We will use the gear train example employed in *Lesson 8* as the test case and as an example for illustrations.

### The Gear Train Example in *Pro/ENGINEER*

The gear train assembly consists of one part and three subassemblies. If you have access to *Pro/ENGINEER*, you may want to open the final assembly, *gear_train_final.asm*, to check the assembled gear train shown in Figure C-2. There are four components in this assembly: *gbox_housing.prt*, *gbox_input.asm*, *gbox_middle.asm*, and *gbox_output.asm*. The input and output gear assemblies consist of one gear each, *Pinion 1* and *Gear 2*, respectively. The middle gear assembly has two gears, *Gear 1* and *Pinion 2*. The four spur gears form two gear pairs: *Pinion 1* and *Gear 1*, and *Pinion 2* and *Gear 2*, as illustrated in Figure C-2. *Gear 1* and *Pinion 2* are mounted on the same shaft.

There are 22 distinct parts in this assembly, as listed in Table C-1. All the parts and assemblies can be found in folder Appendix C.

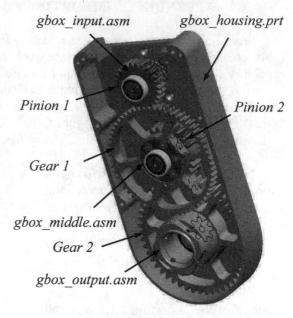

*gbox_input.asm*        *gbox_housing.prt*

*Pinion 1*        *Pinion 2*

*Gear 1*

*gbox_middle.asm*

*Gear 2*

*gbox_output.asm*

Figure C-2  The Gear Train Assembly

Table C-1  List of Parts and Assemblies in Appendix C Folder

| Part/Subassemblies | Part Names | Remarks |
|---|---|---|
| gbox_housing.prt | | |
| gbox_input.asm | wheel_gbox_shaft_input.prt | |
| | wheel_gbox_pinion_1s.prt | Pinion 1 |
| | spacer_12×18×5mm.prt | |
| | spacer_12×20×1mm.prt | |
| | bearing_12×18×8mm.prt (2) | |
| | spacer_10×18×014mm.prt | |
| | wheel_gbox_sft_mid_washer.prt | |
| | screw_tapper_head_5×15.prt | |
| | screw_set_tip_6×6.prt (2) | |
| gbox_middle.asm | wheel_gbox_pinion_2s.prt | Pinion 2 |
| | wheel_gbox_gear_1s.prt | Gear 1 |
| | wheel_gbox_shaft_mid_pinion.prt | |
| | wheel_gbox_shaft_mid_gear.prt | |
| | bearing_12×18×8mm.prt (2) | |
| | screw_tapper_head_5×28.prt (6) | |
| | wheel_gbox_sft_mid_washer.prt (2) | |
| | screw_tapper_head_5×15.prt (2) | |
| | align_pin_4×27mm.prt (2) | |
| gbox_output.asm | wheel_gbox_gear_2s.prt | Gear 2 |
| | wheel_gbox_connect_wheel.prt | |
| | bear_tap_roller25×47×15mm.prt | |
| | screw_straight_head_4×15.prt (10) | |
| | align_pin_4×20mm.prt (2) | |
| | wheel_gbox_connect_wh_setscrew.prt (4) | |

### Importing *Pro/ENGINEER* Parts

We will import the gear housing (*gbox_housing.prt*) shown in Figure C-3 using both options. We will try the first option; i.e., importing solid features. If you have access to *Pro/ENGINEER*, you may choose from the pull-down menu *Tools > Model Player* to see the sequence of feature creation. As shown in Figure C-3 (*Pro/ENGINEER Model Tree* window), there are 8 datum features and 14 solid features. *SOLIDWORKS* will try to import these 14 solid features from *Pro/ENGINEER*.

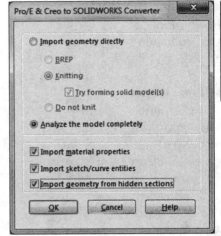

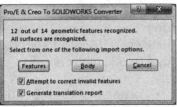

Figure C-5

Figure C-4

Figure C-3  Gear Housing Part
in *Pro/ENGINEER*

*Option 1: Importing Solid Features*

Start *SOLIDWORKS* and choose *File > Open*. Change to folder *Appendix C* (or the folder where you have these gear train parts and assemblies). In the *File Open* dialog box (Figure C-1), pull down the *Files of type* list, and choose *ProE Part (\*prt, , \*.prt.\*, \*.xpr)*. You should see a list of *Pro/ENGINEER* parts in the dialog box. Click *gbox_housing.prt*, and click *Open*. In the *Pro/E & Creo to SOLIDWORKS Converter* dialog box (Figure C-4), choose *Analyze the model completely* (default), and click *Import material properties*, *Import sketch/curve entities*, and *Import geometry from hidden sections*. Click *OK*.

**Translation Report**

Features 14, Recognized 12, Imported 12

| Feature | Recognition st... | Import status |
|---|---|---|
| 1. Extrude1 | Yes | Yes |
| 2. Extrude2 | Yes | Yes |
| 3. Extrude3 | Yes | Yes |
| 4. Extrude4 | Yes | Yes |
| 5. Cut-Revolve1 | Yes | Yes |
| 6. Cut-Revolve2 | Yes | Yes |
| 7. Revolve1 | Yes | Yes |
| 8. Sketch8 | NO | - |
| 9. Extrude5 | Yes | Yes |
| 10. Extrude6 | Yes | Yes |
| 11. _Chamfer | NO | - |
| 12. Extrude7 | Yes | Yes |
| 13. Fillet1 | Yes | Yes |
| 14. Extrude8 | Yes | Yes |
| | 12/14 | 12*/14 (*0 with errors) |

Figure C-6

In the next dialog box (Figure C-5), click the *Features* button. The conversion process will start. You will see sketches and solid features appear in the graphics screen. After about a minute, the *Translation Report* dialog box, as shown in Figure C-6, appears, summarizing the results of the translation. The report indicates that 12 out of 14 features were recognized and translated. The converted model and features listed in the browser are shown in Figure C-7.

As shown in Figure C-7, there is one dangling sketch, *Sketch8*, representing the unrecognizable solid feature in addition to the chamfer feature. You may identify the sketch in the graphics area by clicking its name listed in the browser. In addition, the back plate (*Extrude1* in the browser) is recognized incorrectly. Certainly, *SOLIDWORKS* is capable of importing some parts correctly and completely, especially when the solid features are relatively simple (but not this gear housing part).

If you take a closer look at any of the successful solid features, for example *Extrude3*, you will see that the sketches (for example *Sketch3* of *Extrude3*) of the solid features do not have complete dimensions. Usually a *(−)* symbol is placed in front of the sketch, indicating that the sketch is not fully defined.

Apparently, this translation is not satisfactory. Unfortunately, this translation represents a typical scenario you will encounter for the majority of the parts. In many cases, it may take only a small effort to repair or re-create wrong or unrecognized solid features. However, when you translate an assembly with many parts, the effort could be substantial.

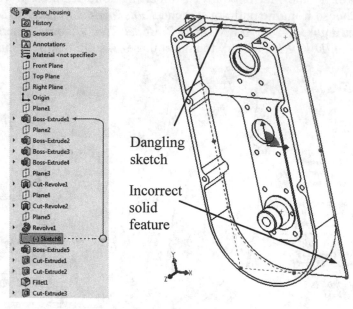

Figure C-7

## Option 2: Importing Geometry

Importing geometry is more straightforward and has a higher successful rate than that of importing solid features.

Repeat the same steps to open the gear housing part *gbox_housing.prt*. In the *Pro/ENGINEER to SOLIDWORKS Converter* dialog box (Figure C-8), choose *Import geometry directly* (default), and then *Knitting* (default) in order to import solid models instead of just surface models. Note that if you choose *BREP* (Boundary Representation), only boundary surfaces will be imported. Click *OK*.

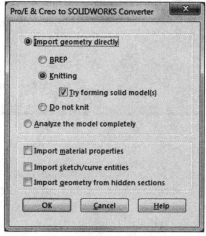

Figure C-8

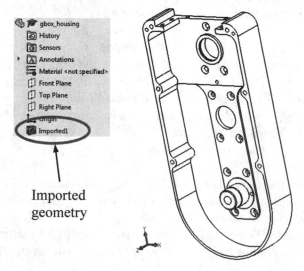

Figure C-9

The conversion process will start. After about a minute or two, the converted model will appear in the graphics screen, as shown in Figure C-9. In addition, an entity *Imported1* will appear in the browser (Figure C-9). As mentioned earlier, there will be no parametric solid feature with dimensions and sketch converted if you choose *Option 2*. However, the geometry converted seems to be accurate. All the geometric features in *Pro/ENGINEER* shown in Figure C-3 were included in this imported feature. This translation is successful. Since we do not anticipate making any change to the gear housing, this imported part is satisfactory. The gear housing part, *gbox_housing.sldprt*, employed in *Lesson 8* was created by using *Option 2*.

### Importing *Pro/ENGINEER* Assembly

We will import the input gear assembly (*gbox_input.asm*) shown in Figure C-10 using both options. We will try *Option 1* first; i.e., importing solid features. As shown in Figure C-10 (*Pro/ENGINEER Model Tree* window), there are 11 parts (and several datum features) in this assembly. *SOLIDWORKS* will try to import this assembly as well as the 11 parts from *Pro/ENGINEER*.

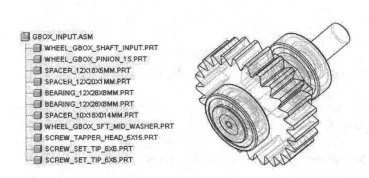

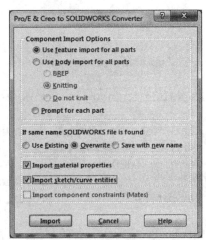

Figure C-10  Input Gear Assembly in *Pro/ENGINEER*

Figure C-11

*Option 1: Importing Solid Features*

Repeat the same steps to open the input gear assembly *gbox_input.asm*. In the *Pro/ENGINEER to SOLIDWORKS Converter* dialog box (Figure C-11), choose *Use feature import for all parts*, and choose *Overwrite* for *If same name SOLIDWORKS file is found* (just in case you have *SOLIDWORKS* files with the same file names in the same folder). Choose *Import material properties* and *Import sketch/curve entities*. Click *Import*. The conversion process will start.

You will see sketches, solid features, and solid models appear in the graphics screen. After about a minute, the translation process is completed. The converted assembly and the browser with parts listed are shown in Figure C-12.

As shown in Figure C-12, parts are not completely converted. Major solid features are missing, for example, the pinion 1 (*wheel_gbox_pinion_1s<1>*), where most solid features are not converted. If you expand the part, you will see that only two extrude features were converted; the remaining entities are mostly sketches. The remaining parts were converted. However, the *Mates* branch in the browser is completely empty, implying that no assembly mates have been imported.

Apparently, this translation is not satisfactory. A non-trivial effort will have to be spent in order to reconstruct the solid features (therefore, solid models) as well as the final assembly.

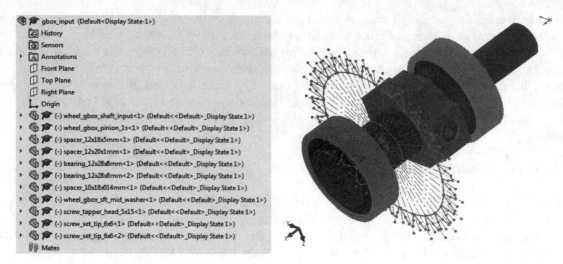

Figure C-12

*Option 2: Importing Geometry*

Importing geometry is also more straightforward for assembly and has a higher rate of success.

Repeat the same steps to open the input gear assembly *gbox_input.asm*. In the *Pro/ENGINEER to SOLIDWORKS Converter* dialog box (Figure C-13), choose *Use body import for all parts* (default), and then *Knitting* (default) in order to import solid models. Choose *Overwrite* for *If same name SOLIDWORKS file is found*, and choose *Import material properties* and *Import sketch/curve entities*. Click *Import*. The conversion process will begin.

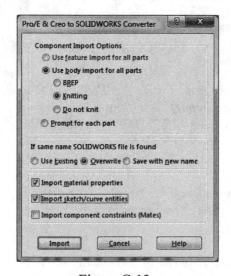

Figure C-13

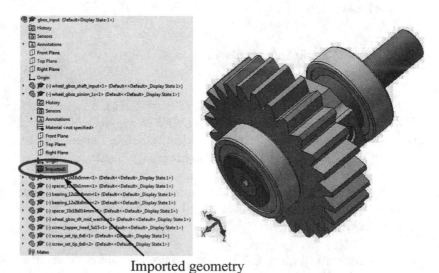

Imported geometry

Figure C-14

After about a minute or two, the converted assembly will appear in the graphics screen, as shown in Figure C-14. The assembly and all 11 parts seem to be correctly imported. If you expand any of the part branch, for example, the gear (*wheel_gbox_pinion_1s<1>*), you will see an imported feature listed, as depicted in Figure C-14. Again, there is no solid feature converted in any of the parts. In addition, the *Mates* branch is empty.

Since we do not anticipate making any change to this input gear assembly, this imported assembly is satisfactory, except it does not have any assembly mates. Assembly of all 11 parts (it may be more for some other cases) will take a non-trivial effort. Since we do not anticipate making changes in how these parts are assembled, we will merge all 11 parts into a single part.

In *SOLIDWORKS*, you can join two or more parts to create a new part in an assembly. The merge operation removes surfaces that intrude into each other's space and merges the parts into a single solid volume. We will insert a new part into the assembly and merge all 11 parts into that new part.

Choose from the pull-down menu *Insert > Component > New Part*. A new part with an assigned name is listed in the browser (Figure C-15). *SOLIDWORKS* is expecting you to select a plane or a flat face to place a sketch for the new part.

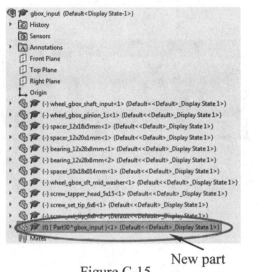

New part

Figure C-15

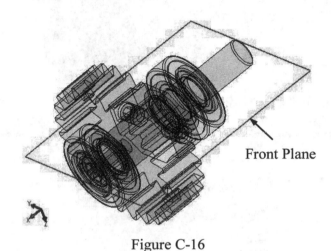

Front Plane

Figure C-16

Click a plane or planar face on a component; for example, pick the assembly *Front* plane from the browser. The *Front* plane will appear in the graphics screen (Figure C-16). In the new part, a sketch opens on the selected plane. Close the sketch by clicking the *Close* button ✖ and choose *Discard Changes and Exit*. Because we are creating a joined part, we do not need a sketch.

Next, we will select all 11 parts and merge them into the new part.

From the browser, click the first part *wheel_box_shaft_input<1>*, press the *Shift* key, and then click the last part, *screw_set_tip_6×6<2>*. All 11 parts will be selected.

From the pull-down menu, choose *Insert > Features > Join*. The *Join* window will appear (overlapping with the browser) as shown in Figure C-17. In the *Join* window, all 11 parts are listed. All you have to do is to click the checkmark on top to accept the parts. Save the part (choosing from the pull-down menu *File > Save*). Choose *Save All* in the *Save Modified Documents* dialog box (Figure C-18), and then click *OK* in the *Save As* dialog box (Figure C-19). A part file *gbox_input.sldprt* is created in the folder. Close the entire assembly model.

Now open the part *gbox_input*. Make sure you open *gbox_input.sldprt* instead of *gbox_input.sldasm*. The part *gbox_input* will appear in the graphics area. In addition, all entities belonging to this part will be listed in the browser, as shown in Figure C-20. Note that there is an arrow symbol ->? to the right of the entity, *Join1*. This symbol indicates that these entities enclosed in this part refer to other parts or assembly. Note that the *Join1* branch has the same symbol. Expand the *Join1* node; you will see 11 parts listed, all with arrows, pointing to the actual parts currently in the same folder.

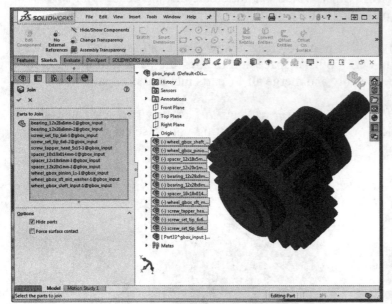

Figure C-17

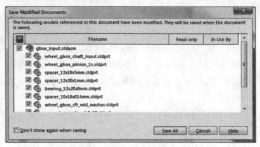

Figure C-18

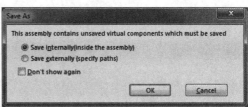

Figure C-19

Join feature

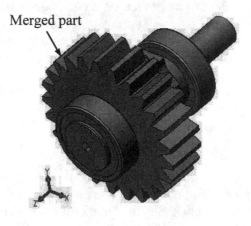

Merged part

Figure C-20

The three gear parts, *gbox_input.sldprt*, *gbox_middle.sldprt*, and *gbox_output.sldprt*, employed for *Lesson 8* were created following the approach discussed. One axis in each part that passes through the center hole of the gear was created simply by intersecting two planes, for example, *Top* and *Right* planes for the axis in *gbox_input.sldprt*, as shown in Figure C-21. These axes are necessary for creating gear pairs, as discussed in *Lesson 8*.

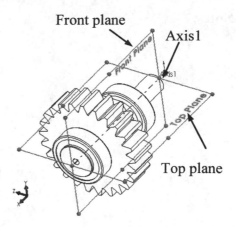

Figure C-21

**Notes:**

**Notes:**

**Notes:**

**Notes:**